Wave Transmission

F. R. Connor

Ph.D., M.Sc., B.Sc.(Eng.) Hons, A.C.G.I.,
C.Eng., M.I.E.E., M.I.E.R.E., M.Inst.P.

Edward Arnold

First published 1972
by Edward Arnold (Publishers) Ltd.,
41 Bedford Square, London WC1B 3DP

Reprinted 1975, 1978, 1980, 1982, 1983

ISBN 0 7131 3278 7

Photoset by The Universities Press, Belfast
and printed in Great Britain by The Pitman Press, Bath.

Preface

This is an introductory book on the important topic of *Wave Transmission*. Electromagnetic waves play an essential part in many communication systems and the book endeavours to present basic ideas concerning the transmission of such waves in a concise and coherent manner. Moreover, to assist in the assimilation of these basic ideas, many worked examples from past examination papers are provided to illustrate clearly the application of the fundamental theory.

The first part of the book deals with the various types of transmission lines used for different applications. A general analysis follows, based on circuit ideas of voltages and currents. Subsequent chapters then consider the alternative concept of fields, which is essential for the proper understanding of waveguide transmission, and the book ends with a useful treatment of microwave theory and techniques.

This book will be found useful by students preparing for London University examinations, degrees of the Council of National Academic Awards, examinations of the Council of Engineering Institutions and for other qualifications such as Higher National Certificates, Higher National Diploma and certain examinations of the City and Guilds of London Institute. It will also be useful to practising engineers in Industry who require a ready source of basic knowledge to help them in their applied work.

Acknowledgements

The author sincerely wishes to thank the Senate of the University of London and the Council of Engineering Institutions for permission to include questions from past examination papers. The solutions and answers provided are his own and he accepts full responsibility for them.

Sincere thanks are also due to the publishers H. W. Peel & Co. Ltd. for permission to reproduce their impedance chart shown in Fig. 20 and also to the City and Guilds of London Institute for permission to include questions from past examination papers. The Institute is in no way responsible for the solutions and answers provided.

Finally, the author wishes to thank the publishers for various useful suggestions and will be grateful to his readers for drawing his attention to any errors in the work.

F.R.C.

Books in this series by the same author:

1. Signals
2. Networks
4. Antennas
5. Modulation
6. Noise

Contents

Symbols used in the book

α	attenuation coefficient
β	phase-change coefficient
γ	propagation coefficient gyromagnetic ratio
δ	skin depth
ρ	specific resistance charge density reflection coefficient
σ	conductivity
λ	wavelength
λ_g	waveguide wavelength
λ_c	cut-off wavelength
ω_0	angular frequency at resonance
c	velocity of light
k_c	the quantity $2\pi/\lambda_c$
J_m	Bessel function of order m
J'_m	first derivative of J_m
s	standing wave ratio reciprocal of skin depth
Z_0	characteristic impedance intrinsic impedance of free space
∇	operator del

Abbreviations used in the book

C.E.I. Part 2	Council of Engineering Institutions examination in Communication Engineering, Part 2.
C. & G.	The City and Guilds of London Institute.
L.U. B.Sc(Eng) Tels.	University of London, B.Sc(Eng). Examination in Telecommunications, Part 3.

1 Introduction

The transmission of energy between a source and some distant point requires the use of a transmission medium generally called a transmission line. In certain cases this requires a physical structure, but in some cases, no structure is required as transmission is achieved directly through free space as an electromagnetic wave. Over the years, several types of transmission lines have been used, each having its own particular applications and limitations, yet finding wide use in various communication systems.

The development of transmission lines springs largely from the use of the familiar two-wire electrical power line for carrying large quantities of power from a generator to its load. However, the demands of communication systems with their far greater frequency requirements, led to the development of various other types of transmission lines which will be treated in this book and the more familiar power line will only be briefly referred to in the text.

Broadly speaking, transmission lines are either lumped lines or distributed lines. Lumped lines are so-called because their electrical parameters such as resistance, inductance and capacitance are lumped at intervals along the line and are unlike the distributed line in which these parameters are uniformly spread over the whole length of line. Most practical lines are of the uniform type as they are easily manufactured and have better characteristics than the lumped line. The most commonly used lines are the two-wire open line, coaxial line, parallel plate or strip line and the waveguide.

1.1 Two-wire open line[1,2]

It consists essentially of two conductors spaced a certain distance apart and used extensively in power systems, telegraphy and telephone systems and in certain areas of radio transmission. Its main use is in the lower frequency range of work and it is simple to manufacture. Such lines are chiefly characterised as having resistance per unit length, while the inductance and capacitance per unit length are usually quite small.

The commonest two-wire line is the overhead power line operating at high voltages. The distance between the conductors is large compared to the conductor diameter, but at high frequencies, radiation losses are minimised by considerably reducing the distance between conductors. Two-wire lines used for communication purposes such as antenna feeders or down-leads to receivers, have characteristic impedances between 70 Ω to 600 Ω. They are spaced apart by dielectric spacers or moulded in some dielectric material. Propagation is essentially as a TEM (transverse electric and magnetic) wave, in which the energy is carried by the fields and the wave is guided along by the conductors as shown in Fig. 1.

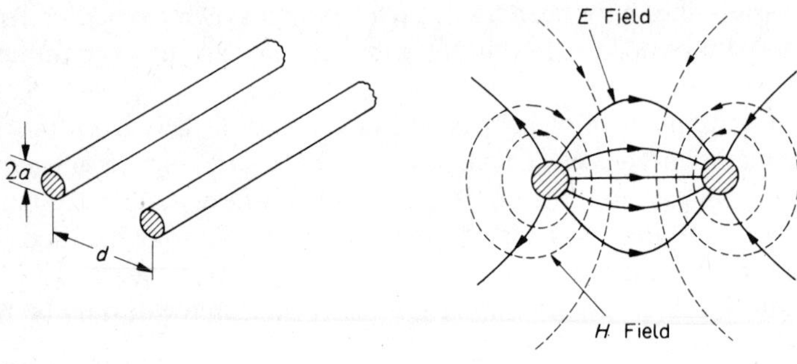

Fig. 1

It can be shown that the relevant parameters, R, L, C per loop metre are given by

$$R_{\text{d.c.}} = \frac{2\rho}{\pi a^2} \text{ ohm} \qquad L = \frac{\mu_0}{\pi} \log_e d/a \text{ H}$$

$$R_{\text{a.c.}} = \frac{1}{\pi a}\left[\frac{\omega\mu\rho}{2}\right]^{1/2} \text{ ohm} \qquad C = \frac{\pi\varepsilon_0}{\log_e d/a} \text{ F}$$

where ρ is the specific resistance, a is the radius of the conductors and d is the distance between their centres. Also

$$Z_0 = \sqrt{\frac{L}{C}} = 276 \log_{10} d/a \text{ ohm}$$

$$v \simeq \frac{1}{\sqrt{\mu_0\varepsilon_0}} = 3 \times 10^8 \text{ m/s}$$

1.2 Coaxial cable[3,4]

The two-wire line is useful mainly at the lower frequencies and up to about 100 MHz in short lengths only. At higher frequencies, serious losses occur due to skin effect in the conductors and radiation from the surface.

Due to the severe radiation losses, a closed field configuration must be used in which an inner conductor is surrounded by an outer cylindrical sheath and is known as a coaxial cable. It has the advantage that the fields are confined within the outer conductor thus eliminating radiation losses and it is also shielded from outside interference. The medium between the conductors may be either air or a dielectric material. Such coaxial cables find extensive use not only at power frequencies where the main problem is one of insulation, but also at very high frequencies for radio and television applications.

A typical air-cored cable has an inner conductor of copper held in position by polythene discs which is surrounded by one or more layers of steel tape for screening or strength. Flexible forms of cable are also used with a polythene dielectric and outer copper braiding for flexibility.

In the field of communications, primary considerations are those of attenuation and distortion at the frequencies of operation, while a secondary consideration is that of power. Although coaxial cables have wide-band capabilities from d.c. up to well into the microwave band, attenuation increases with frequency and so coaxial cables may be designed to operate over definite frequency bands only such as audio, radio or video. However, their large bandwidth can be exploited for multichannel operation whereby several frequencies are sent down the same coaxial cable.

Coaxial cables normally use the TEM mode of propagation as shown in Fig. 2. To ensure that other modes do not exist, the cable size has to be

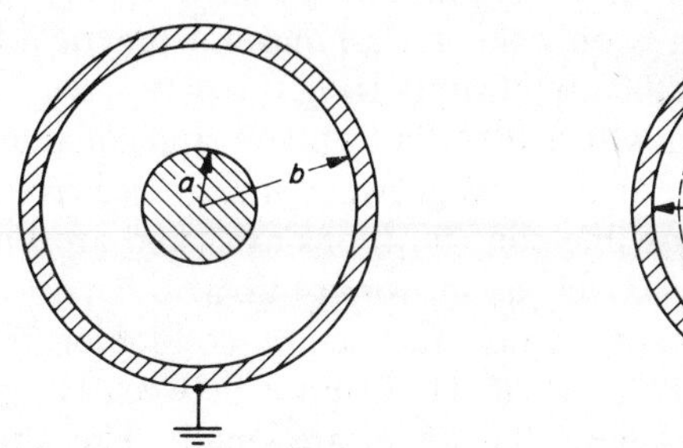

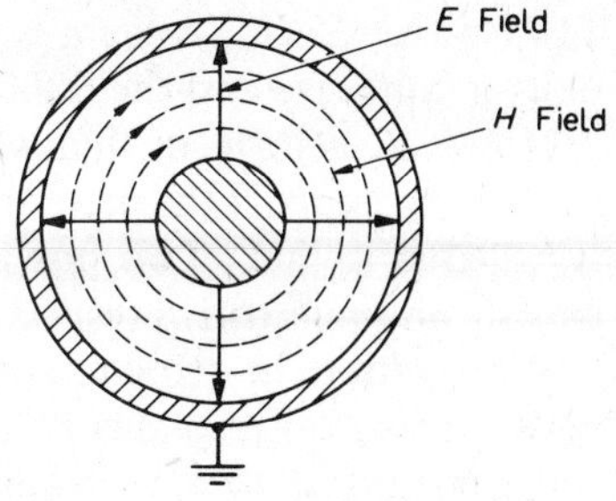

Fig. 2

decreased as the frequency increases, thus reducing its power handling capacity.

A typical coaxial cable is characterised by R, L, G and C per unit length and its relevant parameters are given by:

$$R_{\text{a.c.}} = \frac{1}{2\pi}\left(\frac{\omega\mu\rho}{2}\right)^{1/2}\left[\frac{1}{a} + \frac{1}{b}\right] \text{ ohm/m}$$

$$L = \frac{\mu_0\mu_r}{2\pi}\log_e b/a \text{ henry/m}$$

$$G = \omega C \tan\delta \text{ siemens/m}$$

$$C = \frac{2\pi\varepsilon_0\varepsilon_r}{\log_e b/a} \text{ farad/m}$$

$$Z_0 = \sqrt{L/C} = \frac{138}{\sqrt{\varepsilon_r}}\log_{10} b/a \text{ ohms}$$

$$v = \frac{1}{\sqrt{\mu\varepsilon}} = \frac{3 \times 10^8}{\sqrt{\mu_r\varepsilon_r}} \text{ m/s}$$

$$\alpha = R/2Z_0 \text{ nepers/m}$$

where b is the inner radius of the outer conductor and a is the outer radius of the inner conductor, ρ is the specific resistance and δ is the loss angle of the capacitance C.

1.3 Strip line[5–7]

A form of transmission line having low losses has been known for a long time. It is the parallel plate line with infinite plates and propagating a TEM wave. Such a system is only theoretical and not practical because of its infinite size and the difficulty of supporting the plates.

However, a form of line which uses finite plates and an intervening medium to support the plates has found growing importance recently and is called strip line. Two possible configurations exist and are known as triplate and microstrip respectively as shown in Fig. 3.

Propagation in triplate, for which the centre conductor is placed between two outer plates is TEM, if the distance between the plates is small compared to a wavelength and losses are small. The alternative microstrip geometry has a narrow conductor supported on a dielectric

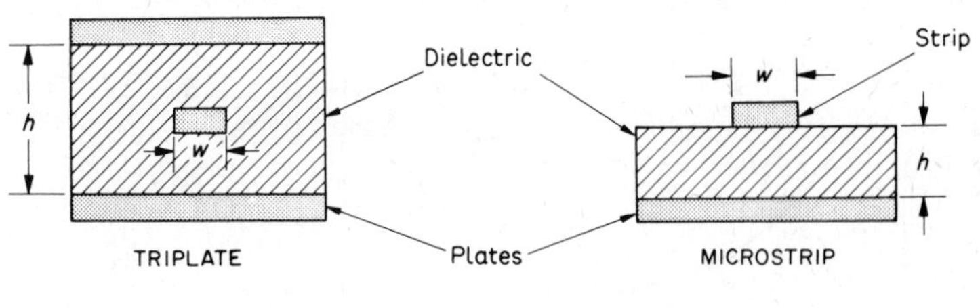

Fig. 3

substrate. It may or may not be covered over by a metal plate serving as an earth plane.

The microstrip geometry has higher losses but is very convenient for manufacturing purposes. Propagation is not true TEM and is somewhat complex though losses can be reduced by using higher dielectric materials. The characteristics of microstrip are difficult to evaluate exactly but depend on the ratio w/h and the dielectric constant ε_r.

Its use has led to the widespread development of microwave integrated circuits since semi-conductor devices can be easily integrated with the microstrip line, by use of a semiconducting material as the substrate. Such circuits are finding wide application at microwave frequencies, both for the manufacture of components such as couplers, circulators, etc., or for complete integrated systems such as receivers.

Microstrip lines are usually fabricated on fibre glass or polystyrene printed circuit boards about 1·5 mm thick with copper lines 3 mm wide. For integrated circuits alumina, silicon or sapphire about 0·25 mm thick are used as substrates with conductors made of copper, aluminium or gold about 0·25 mm wide.

Strip line parameters are usually designed using computerised techniques and yield results for the characteristic impedance Z_0, wavelength λ_{TEM} and attenuation α which are given by

$$Z_0 = \sqrt{\frac{L}{C}} = \frac{377}{\sqrt{\varepsilon_r}}\left(\frac{h}{w}\right) \text{ ohms} \qquad \alpha = \frac{R}{2Z_0} + \frac{gZ_0}{2} \text{ nepers/m}$$

$$\lambda_{TEM} = \frac{\lambda_0}{\sqrt{\varepsilon_r}}$$

where λ_0 is the free space wavelength, ε_r is the relative permittivity of the substrate, R is the effective series resistance per metre of the conductor and ground plane, while g is an effective conductance representing the

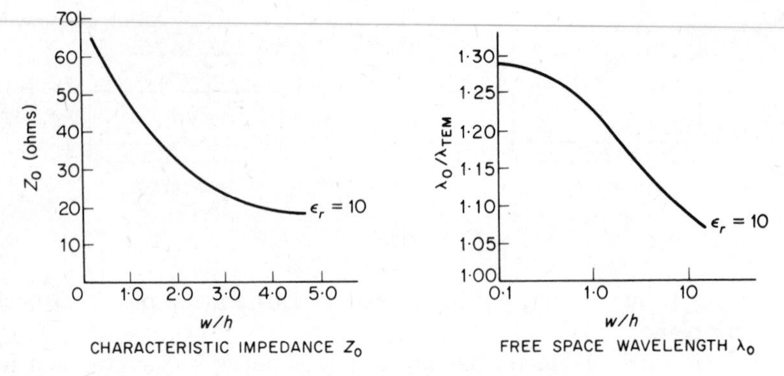

Fig. 4

total dielectric losses. Typical graphs of the relevant parameters are shown in Fig. 4 for various values of w/h and $\varepsilon_r = 10$.

1.4 Waveguides

For minimum losses and high-power transmission at microwave frequencies, the waveguide is used extensively. It consists essentially of a single metallic conductor in the shape of a rectangular box or cylinder, down which electromagnetic waves are propagated. Such guided waves have field configurations somewhat different from those of the previous lines considered and are called either transverse electric (TE) or transverse magnetic (TM) waves. Alternatively, they are known as *H* waves or *E* waves respectively.

Rectangular waveguides have less losses than circular waveguides and are less prone to mode changing. Hence, they are more commonly used but recently the extremely low loss of one of the circular guide modes, the TE_{01} (H_{01} wave), is being exploited for long distance communication purposes.[8,9] Previous problems associated with mode changing and distortion due to phase-change along the guide, are being overcome and it is likely that this will lead to its greater use at millimetric wavelengths, because losses are greatly reduced at these high frequencies. Moreover, waveguide systems provide wideband capabilities and will be used to meet the demands for multichannel circuits.

Waveguides are usually made of brass, copper or aluminium in various standard sizes corresponding to the frequencies used. To reduce losses, the inside walls are sometimes coated with a thin layer of silver or gold. Typical standard sizes for rectangular waveguides are given in Table 1.

Table 1

WG	*Inside Dimensions* (cm)	*Frequency Range* (GHz)	*Band*
6	16·5 × 8·25	1·15–1·72	L
10	7·2 × 3·4	2·60–3·95	S
16	2·28 × 1·01	8·20–12·4	X
22	0·71 × 0·356	26·5–40	Q
26	0·31 × 0·155	60–90	O

At the lower microwave frequencies, hard drawn tubes are usually manufactured but at millimetric wavelengths special electroforming techniques are used for high precision. Waveguide sections are usually coupled together by flanged assemblies which are bolted together and supported periodically on metal stands.

The parameters associated with the rectangular or circular waveguide shown in Fig. 23 may be easily derived from theory using Maxwell's equations and are well established in practice. The important parameters of impedance Z and cut-off wavelength λ_c are closely related to the dimensions of the waveguide, while attenuation losses depend on these factors as well as on the inner surface finish and metal of the waveguide walls. It is easily shown in Section 5.5 that typical values for Z_{TE} and λ_c, for the TE_{10} wave in rectangular waveguide are

$$Z_{TE} = 377\lambda_g/\lambda \text{ ohms}$$
$$\lambda_c = 2a$$

where λ_g is the guide wavelength, λ is the free space wavelength and a is the width of the waveguide.

2
Analysis of lines

Transmission lines used for communication purposes must operate over a range of frequencies and their behaviour is analysed in terms of resistance R, inductance L, conductance G and capacitance C, all defined per unit length of line. For the purposes of analysis, a line of infinite length may be considered and is known as an infinite line. Results obtained for such an infinite line can then be easily applied to the shorter, general line.

2.1 The general line

Consider the general two-wire line shown in Fig. 5 with primary line constants R, L, G and C defined per loop metre (one metre along each wire). If the voltage and current at the input to a short section of length δx are v and i, respectively, there is a drop of voltage across the section and leakage current between the lines. The corresponding values of voltage and current at the output of the section are given by using partial differentiation since v and i are alternating quantities and also vary with distance x along the line.

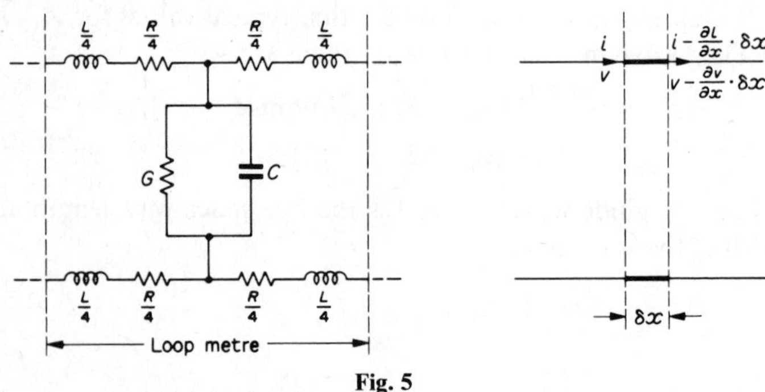

Fig. 5

Hence, we obtain the equations

$$-\frac{\partial v}{\partial x}\,\delta x = R\,\delta x i + L\,\delta x\,\frac{\partial i}{\partial t}$$

$$-\frac{\partial i}{\partial x}\,\delta x = G\,\delta x v + C\,\delta x\,\frac{\partial v}{\partial t}.$$

or
$$-\frac{\partial v}{\partial x} = Ri + L\frac{\partial i}{\partial t} \tag{1}$$

$$-\frac{\partial i}{\partial x} = Gv + C\frac{\partial v}{\partial t} \tag{2}$$

A single sinusoidal voltage or current can be represented by the phasors

$$v = V\mathrm{e}^{\mathrm{j}\omega t}$$
$$i = I\mathrm{e}^{\mathrm{j}\omega t}$$

where V, I are functions of x only. Hence

$$\frac{\partial v}{\partial x} = \frac{\mathrm{d}V}{\mathrm{d}x}\mathrm{e}^{\mathrm{j}\omega t} \qquad \frac{\partial i}{\partial x} = \frac{\mathrm{d}I}{\mathrm{d}x}\mathrm{e}^{\mathrm{j}\omega t}$$

$$\frac{\partial v}{\partial t} = \mathrm{j}\omega V\mathrm{e}^{\mathrm{j}\omega t} \qquad \frac{\partial i}{\partial t} = \mathrm{j}\omega I\mathrm{e}^{\mathrm{j}\omega t}$$

Substituting these results in equations (1) and (2) yields

$$-\frac{\mathrm{d}V}{\mathrm{d}x} = (R + \mathrm{j}\omega L)I$$

$$-\frac{\mathrm{d}I}{\mathrm{d}x} = (G + \mathrm{j}\omega C)V$$

and on further differentiation we obtain

$$\frac{\mathrm{d}^2V}{\mathrm{d}x^2} = (R + \mathrm{j}\omega L)(G + \mathrm{j}\omega C)V$$

$$\frac{\mathrm{d}^2I}{\mathrm{d}x^2} = (R + \mathrm{j}\omega L)(G + \mathrm{j}\omega C)I$$

or
$$\frac{\mathrm{d}^2V}{\mathrm{d}x^2} = \gamma^2 V \tag{3}$$

$$\frac{\mathrm{d}^2I}{\mathrm{d}x^2} = \gamma^2 I \tag{4}$$

where $\gamma = \sqrt{(R + \mathrm{j}\omega L)(G + \mathrm{j}\omega C)}$ is complex and is called the propagation coefficient. It may be represented by

$$\gamma = \alpha + \mathrm{j}\beta$$

where α is the attenuation coefficient and β is the phase-change coefficient.
It is easily shown that

$$\alpha^2 - \beta^2 = RG - \omega^2 LC$$

and

$$2\alpha\beta = \omega(LG + RC)$$

2.2 Secondary line constants

The attenuation constant α accounts for the loss of voltage down the line, while β produces a regular phase shift along the line. This is so, because the wave requires finite time to travel down the line. The quantities α and β are called the secondary constants but are not really constant, as they vary with frequency. A further quantity of interest is the velocity with which energy travels down the line and it is directly related to ω and β.

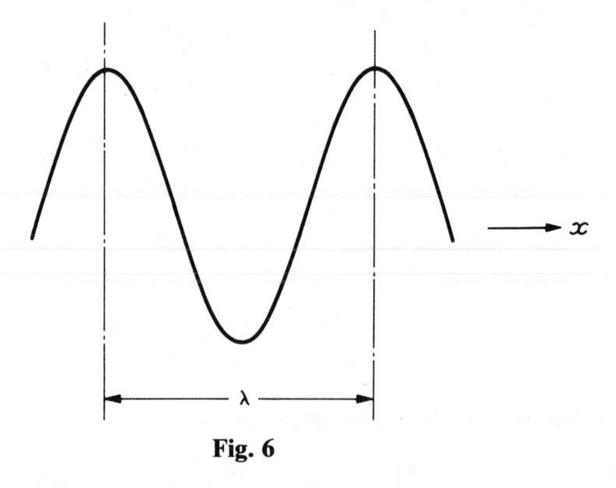

Fig. 6

Consider the wave shown in Fig. 6 as travelling down the line with velocity v and angular frequency ω. The distribution of voltage over a distance $x = \lambda$, will correspond to one full cycle of variation as shown in Fig. 6 and the corresponding phase shift is equal to 2π radians.

Hence

$$\beta\lambda = 2\pi$$

or

$$\beta = \frac{2\pi}{\lambda}$$

Now $v = f\lambda$ where f is the frequency of operation. Hence

$$v = f\frac{2\pi}{\beta} = \frac{\omega}{\beta} \text{ metres/s}$$

Equations (3) and (4) above are second-order differential equations. They are of standard form and by differentiation and substitution, the solutions are readily shown to be

$$V = Ae^{-\gamma x} + Be^{\gamma x}$$

$$I = \frac{A}{Z_0}e^{-\gamma x} - \frac{B}{Z_0}e^{\gamma x}$$

where $Z_0 = \sqrt{(R + j\omega L)/(G + j\omega C)}$ is called the characteristic impedance of the line and A, B are constants determined from the boundary conditions.

2.3 Infinite line

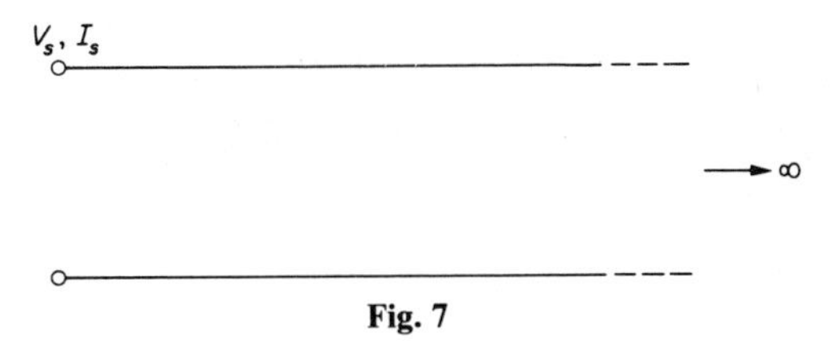

Fig. 7

For the infinite line shown in Fig. 7, let the sending end voltage and current be V_s, I_s respectively.

At $x = 0$, we have

$$V = V_s = Ae^0 + Be^0$$

or

$$V_s = A + B$$

As x tends to infinity, $V \to 0$ since the line voltage is completely attenuated.

Hence
$$0 = Ae^{-\gamma x} + Be^{\gamma x}$$

with $Ae^{-\gamma x} \to 0$ as x tends to infinity and the only possible solution is $B = 0$ with $V_s = A + 0 = A$.

Hence
$$V = V_s e^{-\gamma x}$$

$$I = \frac{V_s}{Z_0}e^{-\gamma x}$$

as the corresponding equations for an infinite line. This leads to the evaluation of Z_0 as

$$\frac{V}{I} = \frac{V_s e^{-\gamma x}}{(V_s/Z_0)e^{-\gamma x}} = Z_0$$

or the ratio of voltage to current at the input to an infinite line or at any point in it is Z_0 and is given the name of characteristic impedance. It depends largely on R, L, G and C which are fixed by the design of a particular line. However, Z_0 also varies with f, the frequency of the wave on the line.

2.4 Hyperbolic solutions

Alternative solutions to the line equations use hyperbolic functions and yield a form useful for solving numerical problems.

We have

$$V(x) = Ae^{-\gamma x} + Be^{\gamma x}$$

$$I(x) = \frac{A}{Z_0} e^{-\gamma x} - \frac{B}{Z_0} e^{\gamma x}$$

Let V_s, I_s, V_r and I_r be the corresponding voltages and currents at the sending end and receiving end respectively.

At $x = 0$, $V = V_s$ and $I = I_s$ with

$$V_s = A + B$$

$$I_s = \frac{A}{Z_0} - \frac{B}{Z_0}$$

or

$$I_s Z_0 = A - B$$

giving

$$A = \frac{V_s + I_s Z_0}{2}$$

$$B = \frac{V_s - I_s Z_0}{2}$$

Substituting into $V(x)$ and $I(x)$ above yields

$$V(x) = e^{-\gamma x}\left[\frac{V_s + I_s Z_0}{2}\right] + e^{\gamma x}\left[\frac{V_s - I_s Z_0}{2}\right]$$

$$I(x) = \frac{e^{-\gamma x}}{Z_0}\left[\frac{V_s + I_s Z_0}{2}\right] - \frac{e^{\gamma x}}{Z_0}\left[\frac{V_s - I_s Z_0}{2}\right]$$

Since $$\cosh \gamma x = \frac{e^{\gamma x} + e^{-\gamma x}}{2}$$

$$\sinh \gamma x = \frac{e^{\gamma x} - e^{-\gamma x}}{2}$$

Hence $$V(x) = V_s \cosh \gamma x - I_s Z_0 \sinh \gamma x$$

$$I(x) = I_s \cosh \gamma x - \frac{V_s}{Z_0} \sinh \gamma x$$

2.5 Practical line

Since the input impedance of an infinite line is a constant and equal to Z_0, shorter practical lines using this property can be employed instead. This is illustrated in Fig. 8 where a short length l of an infinite line is cut away and the remainder which is still an infinite line whose input impedance is Z_0, is replaced by a lumped component of value Z_0, as shown on the right.

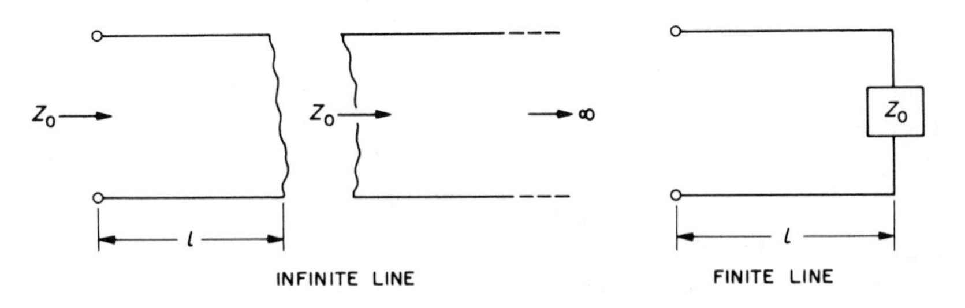

Fig. 8

A finite length of line which is terminated in Z_0 is called a correctly terminated or matched line and it has properties similar to an infinite line. Hence, the equations derived for the infinite line also hold in this case.

2.6 General termination

In many cases, the practical line may have a general termination Z_R and it is necessary to evaluate the input impedance Z_s.

Let the input impedance of the line shown in Fig. 9 be Z_s. Using the

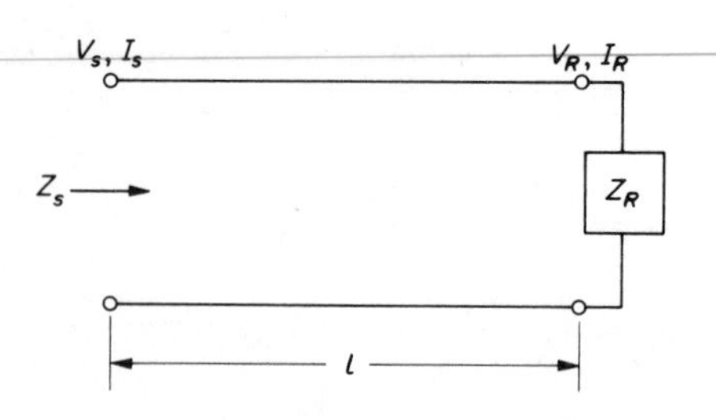

Fig. 9

quantities given in Fig. 9 we obtain

$$Z_s = V_s/I_s$$

$$Z_R = V_R/I_R$$

At $x = l$

$$V_R = V_s \cosh \gamma l - I_s Z_0 \sinh \gamma l$$

$$I_R = I_s \cosh \gamma l - V_s/Z_0 \sinh \gamma l$$

or

$$Z_R = \frac{V_R}{I_R} = \frac{V_s \cosh \gamma l - I_s Z_0 \sinh \gamma l}{I_s \cosh \gamma l - V_s/Z_0 \sinh \gamma l}$$

Substituting for V_s from above, in terms of Z_s and I_s yields

$$\frac{V_s}{I_s} = Z_s = Z_0\left[\frac{Z_R \cosh \gamma l + Z_0 \sinh \gamma l}{Z_0 \cosh \gamma l + Z_R \sinh \gamma l}\right]$$

or

$$Z_s = Z_0\left[\frac{Z_R/Z_0 + \tanh \gamma l}{1 + Z_R/Z_0 \tanh \gamma l}\right]$$

which shows that Z_s is in general complex.

For high frequency lines operating around 100 MHz, α is small compared to β and so $\cosh \gamma l \simeq \cos \beta l$ and $\sinh \gamma l \simeq \text{j} \sin \beta l$.

Hence

$$Z_s = Z_0\left[\frac{Z_R \cos \beta l + \text{j}Z_0 \sin \beta l}{Z_0 \cos \beta l + \text{j}Z_R \sin \beta l}\right]$$

or

$$Z_s = Z_0\left[\frac{Z_R + \text{j}Z_0 \tan \beta l}{Z_0 + \text{j}Z_R \tan \beta l}\right]$$

2.7 Special cases

Two special cases are the open-circuit line impedance and short-circuit line impedance, for the line of length l shown in Fig. 10.

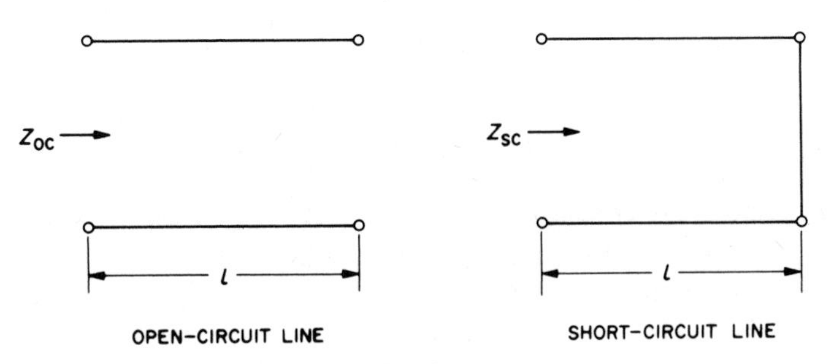

Fig. 10

(a) Open-circuit line

Here $Z_R = \infty$, $Z_s = Z_{oc}$. Hence

$$Z_{oc} = Z_0\left[\frac{\cosh \gamma l + Z_0/Z_R \sinh \gamma l}{Z_0/Z_R \cosh \gamma l + \sinh \gamma l}\right]$$

or
$$Z_{oc} = Z_0 \coth \gamma l$$

(b) Short-circuit line

Here $Z_R = 0$, $Z_s = Z_{sc}$. Hence

$$Z_{sc} = Z_0\left[\frac{Z_R \cosh \gamma l + Z_0 \sinh \gamma l}{Z_0 \cosh \gamma l + Z_R \sinh \gamma l}\right]$$

or
$$Z_{sc} = Z_0 \tanh \gamma l$$

Two further equations can be obtained from the last two results. These are:

$$Z_0 = \sqrt{Z_{oc} \,.\, Z_{sc}}$$

and
$$\tanh \gamma l = \sqrt{\frac{Z_{sc}}{Z_{oc}}}$$

from which Z_0 and γ for a line can be obtained.

EXAMPLE 1

Derive an expression for the input impedance of a loss-free transmission line of length l, terminated by an impedance Z. Assume the voltage and current at distance x from the receiving end to be given by

$$V = Ae^{j\beta x} + Be^{-j\beta x}$$
$$IZ_0 = Ae^{j\beta x} - Be^{-j\beta x}$$

where A and B are constants, $\beta = \omega\sqrt{LC}$ is the phase-change coefficient and $Z_0 = \sqrt{L/C}$ is the characteristic impedance, L and C are the inductance and capacitance per unit length of line and ω is the angular frequency.

If $L = 0{\cdot}60\ \mu H/m$, $C = 240$ pF/m and $\omega = 2\pi \times 10^8$ rad/s, determine (1) the phase-change coefficient β and wavelength λ in the line and (2) the input impedance for a line of length $l = \lambda/4$ terminated by an impedance $Z = -j100\ \Omega$. (L.U. El. Th. & Meas. Part 2, 1969)

Solution

The input impedance has been derived in Section 2.6 where x was measured from the sending end. The expressions for V and I given in the question will correspond with those in the text if x is replaced by $-x$.

PROBLEM

$$\beta = \omega\sqrt{LC}$$
$$= 2\pi \times 10^8\sqrt{6 \times 10^{-7} \times 240 \times 10^{-12}}$$

or $$\beta = 2{\cdot}4\pi$$

Also $$\lambda = \frac{2\pi}{\beta} = \frac{2\pi}{2{\cdot}4\pi} = 0{\cdot}833 \text{ m}$$

and $$Z_0 = \sqrt{\frac{L}{C}} = \left[\frac{6 \times 10^{-7}}{240 \times 10^{-12}}\right]^{1/2} = 50\ \Omega$$

Now $$Z_{in} = Z_0\left[\frac{Z_R \cos\beta l + jZ_0 \sin\beta l}{Z_0 \cos\beta l + jZ_R \sin\beta l}\right]$$

where $$Z_R = -j100$$

$$\beta l = \frac{2\pi}{\lambda}\cdot\frac{\lambda}{4} = \frac{\pi}{2}$$

Hence $$\cos\beta l = \cos\pi/2 = 0$$
$$\sin\beta l = \sin\pi/2 = 1$$

with
$$Z_{in} = Z_0\left[\frac{jZ_0}{jZ_R}\right] = Z_0^2/Z_R$$
$$= \frac{(50)^2}{-j100}$$

or
$$Z_{in} = j25\,\Omega$$

2.8 Line classification

Transmission lines are used over a wide range of frequencies and have certain particular characteristics which need further consideration.

(a) Loss-free line

In this case $R = 0$, $G = 0$. Here

$$\gamma = \sqrt{j\omega L \times j\omega C} = j\omega\sqrt{LC}$$

Equating real and imaginary parts yields

$$\alpha = 0$$
$$\beta = \omega\sqrt{LC}$$

Also
$$Z = \sqrt{\frac{j\omega L}{j\omega C}} = \sqrt{\frac{L}{C}}$$

which is a pure resistance.

This is a rather ideal case and the nearest practical example is the low-loss line.

(b) Low loss line

Here $G \simeq 0$ and R is small where $R \ll \omega L$. Hence

$$\gamma = \sqrt{(R + j\omega L)(j\omega C)} = j\omega\sqrt{(R/j\omega + L)C}$$
$$= j\omega\sqrt{LC(1 + R/j\omega L)}$$
$$= j\omega\sqrt{LC}\left[1 - j\frac{R}{\omega L}\right]^{1/2}$$
$$= j\omega\sqrt{LC}\left[1 - j\frac{R}{2\omega L}\right]$$

by the Binomial theorem.

Hence $$\gamma = \frac{\omega R\sqrt{LC}}{2\omega L} + j\omega\sqrt{LC}$$

$$= \frac{R\sqrt{LC}}{2L} + j\omega\sqrt{LC}$$

$$= \frac{R}{2}\sqrt{\frac{C}{L}} + j\omega\sqrt{LC}$$

giving $$\alpha = \frac{R}{2}\sqrt{\frac{C}{L}}$$

$$\beta = \omega\sqrt{LC}$$

Also $$Z_0 = \sqrt{\frac{R + j\omega L}{j\omega C}} = \sqrt{\frac{L}{C}[1 - j(R/\omega L)]}$$

$$\simeq \sqrt{\frac{L}{C}}[1 - R/\omega l]^{1/2}$$

or $$Z_0 \simeq \sqrt{\frac{L}{C}}[1 - j(R/2\omega L)] \simeq \sqrt{\frac{L}{C}} - j\frac{R}{2\omega}\sqrt{\frac{1}{LC}}$$

This reduces to $\sqrt{L/C}$, a pure resistance at the higher frequencies, due to the large value of ω in the denominator of the second term.

Hence $$\alpha \simeq R/2Z_0$$

$$\beta = \omega\sqrt{LC}$$

If $G \neq 0$, then $$\alpha \simeq R/2Z_0 + GZ_0/2$$

(c) Low frequency line

Typically, it is an audio frequency telephone line and G and L may be low.

Hence $$j\omega L \ll R$$

$$G \ll j\omega C$$

with $$\gamma = \sqrt{R(j\omega C)}$$
$$= \sqrt{j\omega RC}$$
$$= \sqrt{\omega RC}\angle 45^\circ$$

or $$\gamma = \sqrt{\omega RC}\cos 45^\circ + j\sqrt{\omega RC}\sin 45^\circ$$

Hence $$\alpha = \beta = \sqrt{\frac{\omega RC}{2}} \quad \text{numerically}$$

Also $$v = \omega/\beta = \sqrt{\frac{2\omega}{RC}}$$

and $$Z_0 = \sqrt{\frac{R}{j\omega C}} = \sqrt{\frac{R}{\omega C}}\angle -45^\circ$$

(d) High frequency line

A two-wire radio-frequency line operating around 100 MHz comes under this classification.

Here $$\omega L \gg R$$
$$\omega C \gg G$$

Hence $$\gamma = \sqrt{j\omega L \times j\omega C} = j\omega\sqrt{LC}$$

giving $$\alpha = 0$$
$$\beta = \omega\sqrt{LC}$$

with $$v = \omega/\beta = 1/\sqrt{LC}$$

As L and C are very small for such lines v tends to c, the velocity of light which equals 3×10^8 m/s.

Also $$Z_0 = \sqrt{\frac{j\omega L}{j\omega C}} = \sqrt{L/C}$$

a pure resistance.

(e) Distortionless line

This type of line is of interest in designing practical lines with little or no distortion. The two types of distortion which are possible are attenuation distortion, in which certain frequencies are severely attenuated and may even disappear and phase distortion, in which different frequencies suffer different amount of phase shift which is likely to be serious, especially in television.

The analysis is based on the fact that α and β are both functions of frequency, as is shown by the following expressions:

$$\alpha^2 = \tfrac{1}{2}[\{(RG + \omega^2LC)^2 + \omega^2(LG - RC)^2\}^{\frac{1}{2}} + (RG - \omega^2LC)]$$

$$\beta^2 = \tfrac{1}{2}[\{(RG + \omega^2LC)^2 + \omega^2(LG - RC)^2\}^{\frac{1}{2}} - (RG - \omega^2LC)]$$

Putting $L/R = C/G \quad \text{or } LG = RC$

yields

$$\alpha \simeq \sqrt{RG}$$

$$\beta \simeq \omega\sqrt{LC}$$

$$v = \omega/\beta = 1/\sqrt{LC} \quad \text{(a constant)}$$

Hence, α is independent of frequency and β is proportional to frequency, the latter amounting to a linear phase characteristic.

The first condition $\alpha \simeq \sqrt{RG}$ yields no attenuation distortion, since all frequencies are attenuated by the same amount. The second condition $\beta \simeq \omega\sqrt{LC}$ yields no phase distortion since the velocity of travel of all frequencies is the same and their relative phase difference is unchanged down the line.

Also
$$Z_0 = \sqrt{\frac{R + j\omega L}{G + j\omega C}} = \sqrt{\frac{L(R/L + j\omega)}{C(G/C + j\omega)}}$$

or
$$Z_0 = \sqrt{L/C} \quad \text{which is a pure resistance.}$$

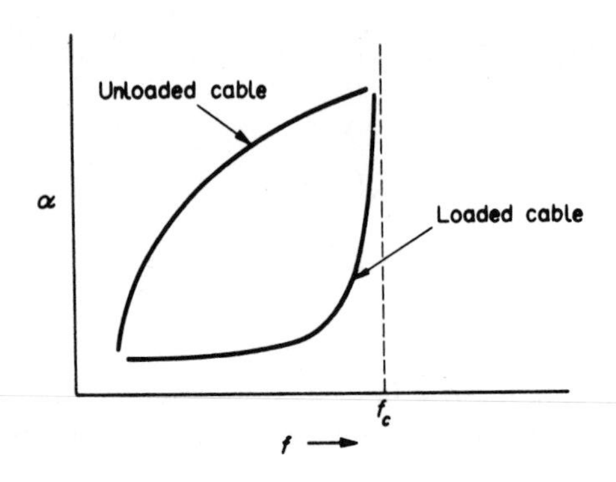

Fig. 11

The condition $LG = RC$ is called the *distortionless condition* and signifies the ideal condition for no attenuation or phase distortion. It is difficult to achieve in a practical line such as a cable since G is small. Hence L is increased artifically at intervals along the line by inserting lumped 'loading coils', a technique known as 'loading'.[10] This practice, which was well established in earlier days, has now been superseded by the use of repeaters, both in underground and submarine cables.

The overall effect of loading was to produce a low-pass filter with a cut-off frequency which had a much lower α over the pass-band than its unloaded counterpart as shown in Fig. 11.

2.9 Phase and group delay[11]

When a single frequency signal travels down a line, it suffers a phase shift of β rad/m. Hence, over a wavelength λ, the phase shift is 2π rads and we obtain

$$\beta\lambda = 2\pi$$

or

$$\beta = 2\pi/\lambda = 2\pi f / f\lambda = \omega/v_p$$

where v_p is defined as the phase velocity of the wave.

Hence $$v_p = \omega/\beta \text{ metres/s.}$$

The reciprocal of v_p has the dimensions of seconds if the distance considered is a metre. It is called the phase delay, or time delay suffered by the wave in travelling unit distance.

Hence $$\text{Phase Delay} = 1/v_p = \beta/\omega \text{ seconds}$$

For no phase distortion to occur, β must be proportioned to ω i.e. a linear phase characteristic. Hence, the corresponding phase delay must be constant and independent of frequency.

Similarly, when two or more frequencies are present on the line as in an A.M. signal, the waves combine to form a group and the peak of the envelope travels forward with a group velocity given by

$$v_g = \mathrm{d}\omega/\mathrm{d}\beta \text{ m/s}$$

The reciprocal of this also has the dimensions of seconds per unit distance and corresponds to a group delay.

Hence $$\text{Group Delay} = 1/v_g = \mathrm{d}\beta/\mathrm{d}\omega \text{ seconds}$$

For the group of frequencies to remain in step as a group, the group delay must also be constant. Hence, $d\beta/d\omega$ must be constant, which amounts to a linear variation of β with ω over the range of frequencies of the group, otherwise distortion will again arise.

Phase or group delay is not serious in audio amplifiers or telephone lines provided it is not too great, as the ear cannot distinguish between small phase differences. However, in television, group delay is very important as it can lead to distortion since the picture brightness depends upon the peak of the video signal, which must not be out of step due to variable group delay. Figure 12 shows a typical plot of group delay in a video cable.

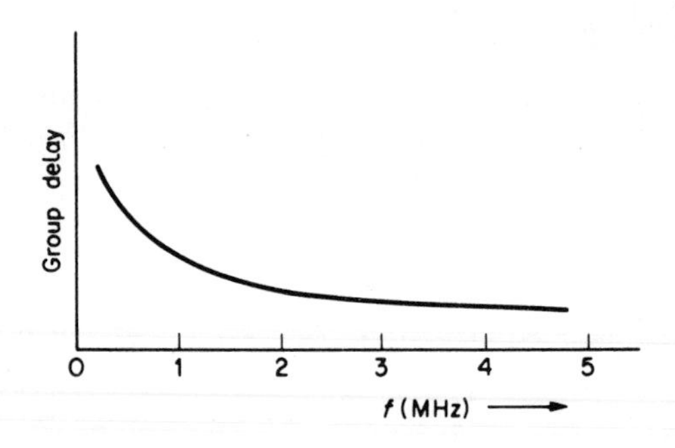

Fig. 12

EXAMPLE 2

A loss-free cable is a quarter of a wavelength long and is excited by a 1·0 V constant-voltage sinusoidal source. The cable is terminated by a resistor of 73·5 Ω, which is not quite equal to the characteristic impedance of the cable. The input current is found to be 15·0 mA. Determine (1) the characteristic impedance of the cable; (2) the phase of the input current relative to that of the voltage source.

(L.U.B.Sc(Eng.) Pt. 2, Elect. Th. & Meas. 1968)

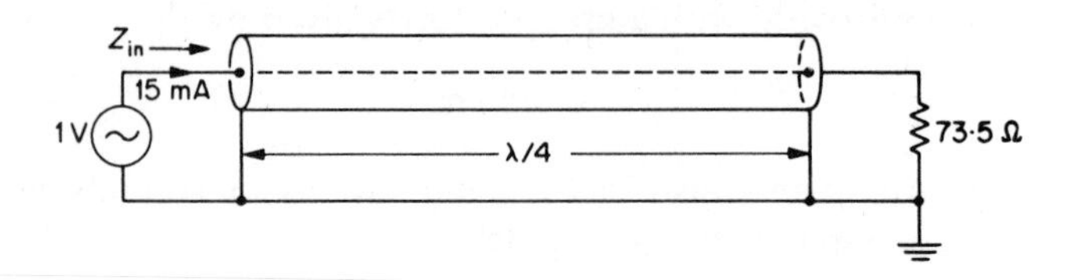

Fig. 13

Solution

$$Z_{in} = Z_0\left[\frac{Z_R \cosh \gamma l + jZ_0 \sinh \gamma l}{Z_0 \cosh \gamma l + jZ_R \sinh \gamma l}\right]$$

$$= Z_0\left[\frac{Z_R + jZ_0 \tanh \gamma l}{Z_0 + jZ_R \tanh \gamma l}\right]$$

Since the cable is loss free, $\alpha = 0$.

Hence $\tanh \gamma l = \dfrac{\sinh \gamma l}{\cosh \gamma l} = \dfrac{j \sin \beta l}{\cos \beta l} = j \tan \beta l$

with $Z_{in} = Z_0\left[\dfrac{Z_R - Z_0 \tan \beta l}{Z_0 - Z_R \tan \beta l}\right] = Z_0\left[\dfrac{Z_R/\tan \beta l - Z_0}{Z_0/\tan \beta l - Z_R}\right]$

Now $\tan \beta l = \tan(2\pi/\lambda \,.\, \lambda/4) = \tan \pi/2 = \infty$

Hence $Z_{in} = Z_0\left[\dfrac{0 - Z_0}{0 - Z_R}\right] = Z_0^2/Z_R$

with $Z_0^2 = Z_{in} \,.\, Z_R = \dfrac{73{\cdot}5}{15 \times 10^{-3}}$

or $Z_0 = 70\,\Omega$ a pure resistance

Since Z_0 and Z_R are pure resistances, $Z_{in} = Z_0^2/Z_R$ is also a pure resistance. It is, in fact, the resistive load of the generator and so the input-voltage and current are in phase.

EXAMPLE 3

Explain what is meant by the terms phase and group velocity.

Derive an expression for the phase coefficient of a transmission line in terms of the resistance R, inductance L, and capacitance C per unit length of line (the leakance is negligible). Hence, obtain expressions for the phase and group velocities in the line. Show that when the quantity Q $(= \omega L/R)$ is large both velocities approach a value of $1/\sqrt{LC}$.

(L.U.B.Sc(Eng.) Tels. Pt. 3, 1969)

Solution

The explanation of phase and group velocities are given in the text in Section 2·9.

The propagation coefficient is given by

$$\begin{aligned}\gamma &= \sqrt{(R + j\omega L)(G + j\omega C)} \\ &= \sqrt{(R + j\omega L)(j\omega C)} \\ &= j\omega\sqrt{(R/j\omega + L)C} \\ &= j\omega\sqrt{(R/j\omega + 1)LC} = j\omega\sqrt{LC}\,[1 - j(R/\omega L)]^{1/2}\end{aligned}$$

Since Q is large, $\omega L \gg R$

Hence $\quad \gamma = j\omega\sqrt{LC}\,[1 - j(R/2\omega L)]$

by the Binomial theorem.

or $$\gamma = \frac{\omega R\sqrt{LC}}{2\omega L} + j\omega\sqrt{LC}$$

Hence $$\alpha = \frac{R}{2}\sqrt{C/L}$$

$$\beta = \omega\sqrt{LC}$$

Now $$v_p = \omega/\beta = \frac{\omega}{\omega\sqrt{LC}} = 1/\sqrt{LC}$$

$$v_g = d\omega/d\beta = \frac{d}{d\beta}(\beta/\sqrt{LC}) = 1/\sqrt{LC}$$

Hence, both v_p and v_g approach the value $1/\sqrt{LC}$ when Q is large.

EXAMPLE 4

The angle of the characteristic impedance of a telephone cable at low speech frequencies approaches $-45°$. Explain why this is the case and show that, as a consequence, the attenuation and phase coefficients are numerically equal at these frequencies.

Such a cable is driven from a low-frequency source having an internal impedance equal to the characteristic impedance of the line and terminated at the far end in a manner firstly to avoid reflection and secondly, to take the maximum power from the line. Distinguish between these conditions and calculate the ratio of the corresponding load powers.

(L.U.B.Sc(Eng.) Tels. Pt. 3, 1967)

Solution

The first part has been solved in Section 2.8(c).

For the second part, let Z_G be the internal impedance of the generator and Z_L that of the load (Fig. 14).

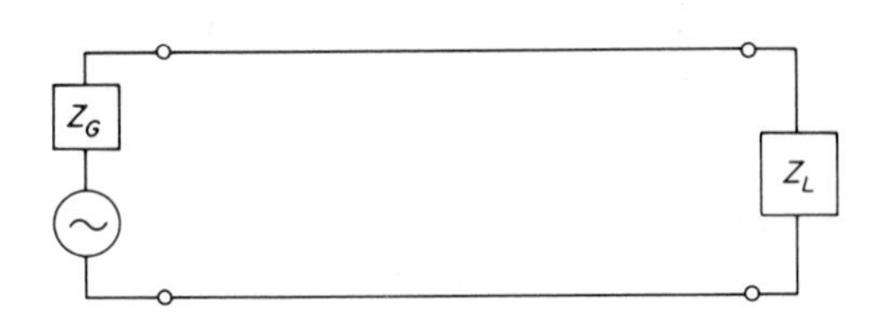

Fig. 14

To avoid reflections, the generator should be matched to the load i.e. the generator and load impedances should have the same magnitude and angle as that of the line.

Hence $$Z_G = Z_L = |Z_0| \underline{/-45^\circ}$$

If I_1 is the current in the load, by Thevenin's theorem

$$I_1 = \frac{V_0}{|Z_0| \underline{/-45^\circ} + |Z_0| \underline{/-45^\circ}} = \frac{V_0 \underline{/45^\circ}}{2\,|Z_0|}$$

where V_0 is the generator voltage on open-circuit.

or $$|I_1| = V_0/2\,|Z_0|$$

The real power P_1 in the load is given by

$$P_1 = |I_1|^2 \times \text{Real part of } |Z_0| \underline{/-45^\circ}$$

$$= \frac{V_0^2}{4\,|Z_0|^2} \times \frac{|Z_0|}{\sqrt{2}} = \frac{V_0^2}{4\sqrt{2}\,|Z_0|}$$

For maximum power transfer, the load impedance must be the same as the line impedance, but of conjugate angle.

Let I_2 be the load current in a load $|Z_0| \underline{/\theta}$ where $\theta = +45^\circ$. Hence

$$I_2 = \frac{V_0}{|Z_0| \underline{/-45^\circ} + |Z_0| \underline{/45^\circ}} = \frac{V_0}{\sqrt{2}\,|Z_0|}$$

The real power P_2 in the load is given by

$$P_2 = |I_2|^2 \times \text{Real part of } |Z_0| \underline{/45^\circ}$$

$$= \frac{V_0^2}{2|Z_0|^2} \times \frac{|Z_0|}{\sqrt{2}} = \frac{V_0^2}{2\sqrt{2}|Z_0|}$$

Hence $$P_1/P_2 = \frac{V_0^2}{4\sqrt{2}|Z_0|} \times \frac{2\sqrt{2}|Z_0|}{V_0^2} = \frac{1}{2}$$

EXAMPLE 5

A uniform telephone cable has the primary coefficients per loop mile $C = 0{\cdot}06\ \mu\text{F}$, $L = 1$ mH, $R = 100\ \Omega$. The conductance is negligible. Calculate the frequency at which the angle of the characteristic impedance is $-22{\cdot}5^\circ$. What additional distributed inductance per mile is required to reduce this frequency to 300 Hz? If this inductance is provided by the insertion of loading coil at regular intervals to produce a nominal cut-off frequency of 4 kHz, estimate (i) the spacing between adjacent coils and (ii) the inductance of each coil.

(L.U.B.Sc(Eng.) Tels. Pt. 3, 1966)

Solution

$$Z_0 = \sqrt{\frac{R + j\omega L}{j\omega C}} = \sqrt{\frac{(R^2 + \omega^2 L^2)^{1/2} \underline{/\tan^{-1} \omega L/R}}{\omega C \underline{/90^\circ}}}$$

Hence $\sqrt{\tan^{-1} \omega L/R - 90^\circ} = -22{\cdot}5^\circ$

or $\frac{1}{2} \tan^{-1} \omega L/R - 45^\circ = -22{\cdot}5$

giving $\tan^{-1} \omega L/R = 45^\circ$

Hence $\omega L/R = 1$

or $\omega = R/L = 100/10^{-3} = 10^5$ rad/s

and $f = 10^5/2\pi = 15{\cdot}9$ kHz

If the total inductance per loop mile is L to reduce the frequency to 300 Hz, we have

$$\tan^{-1} \frac{2\pi \times 300 \times L}{100} = 45^\circ$$

or $$2\pi \times 3 \times L = 1$$

giving $$L = 53 \text{ mH per loop mile}$$

Hence, additional inductance L' required *per loop mile* is

$$L' = (53 - 1) \text{ mH} = 52 \text{ mH}$$

With loading coils, the loaded cable behaves as a low-pass filter. Let d be the distance between the loading coils, L_c is the coil inductance per *loop* mile, L is the line inductance per *loop* mile, and C is the capitance per *loop* mile.

The cut-off frequency f_c is given by

$$f_c = \frac{1}{\pi\sqrt{(L_c + L)dCd}}$$

or $$(L_c + L)Cd^2 = 1/\pi^2 fc^2$$

Substituting the values $L = 1$ mH, $L_c = 52$ mH and $C = 0{\cdot}016\ \mu$F, all per loop mile, yields

$$d^2 = \frac{1}{\pi^2 \times 16 \times 10^6 \times 53 \times 10^{-3} \times 0{\cdot}06 \times 10^{-6}}$$

or $$d = 1{\cdot}41 \text{ miles}$$

Hence $$L_{\text{coil}} = 52 \times 10^{-3} \times 1{\cdot}41$$

or $$L_{\text{coil}} = 73{\cdot}5 \text{ mH}$$

3
Reflections on lines

If the terminating impedance on a transmission line is not Z_0, then a discontinuity exists at the end of the line and reflected waves of voltage and current will be set up on the line and they travel back towards the sending end. As far as energy is concerned, this means that some of the incident energy is reflected back, while the rest is absorbed in the load. The extreme cases of reflection occur on open-circuit and short-circuit.

3.1 Open-circuited line

When the incident wave reaches the open circuit at the end of the line, the magnetic field collapses since the current is reduced to zero. This induces a voltage on the line which adds to the existing voltage on the line and equals it in magnitude. Hence, there is a *voltage doubling* effect.

The induced voltage on the line then travels back along the line and may be absorbed by the generator impedance if it equals Z_0, the characteristic impedance of the line. If not, it travels to and fro along the line till it is finally attenuated completely.

3.2 Short-circuited line

Since the voltage is zero at the end of the line, there must be a phase reversal of the incident voltage at the end of the line so that the incident and reflected waves cancel one another at the end of the line. The reflected wave then travels back along the line and is either absorbed by the generator impedance or completely attenuated, as in the previous case.

Hence, in both these cases, there are two waves on the line, (a) the incident wave, and (b) the reflected wave. The resultant wave at any part of the line is the algebraic sum of the two waves and it produces a *standing wave* on the line.

The incident and reflected waves travel along the line and hence are called *travelling waves*. The resultant standing wave pattern on the line varies in magnitude, but remains *stationary* and contains maxima and minima at certain points on the line, where adjacent maxima or minima are separated by $\lambda/2$. A diagram illustrating these various points for two

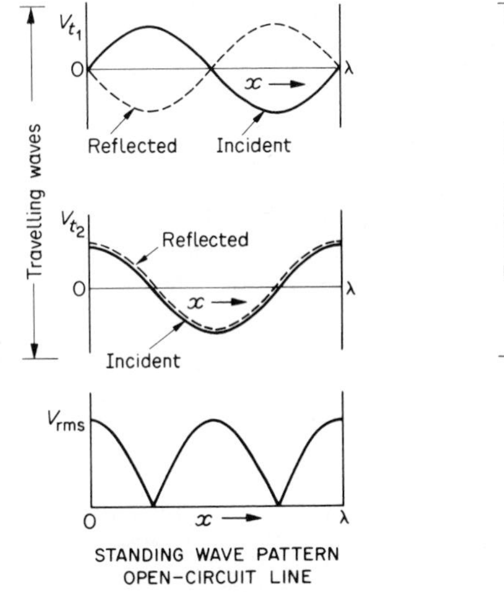

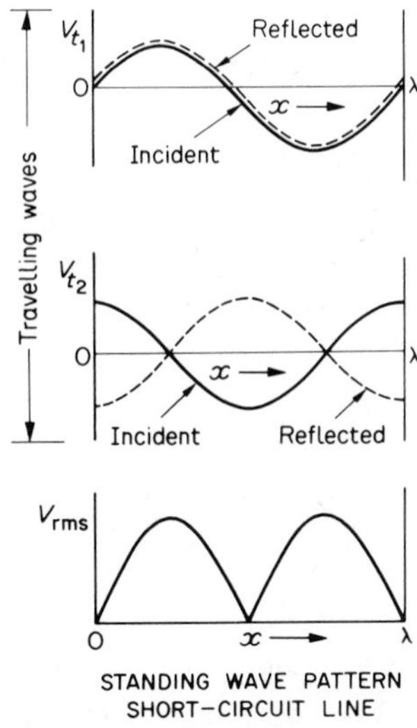

Fig. 15

different instances of time is shown in Fig. 15, in which the line is assumed to be lossless and of length λ. The r.m.s. value of the voltage and current at various points on the line are shown in Fig. 16.

The ratio of voltage to current waves gives the impedance at the various points on the line. This is shown in Fig. 16 and for convenience, distances are measured from the receiving end. In particular, it will be

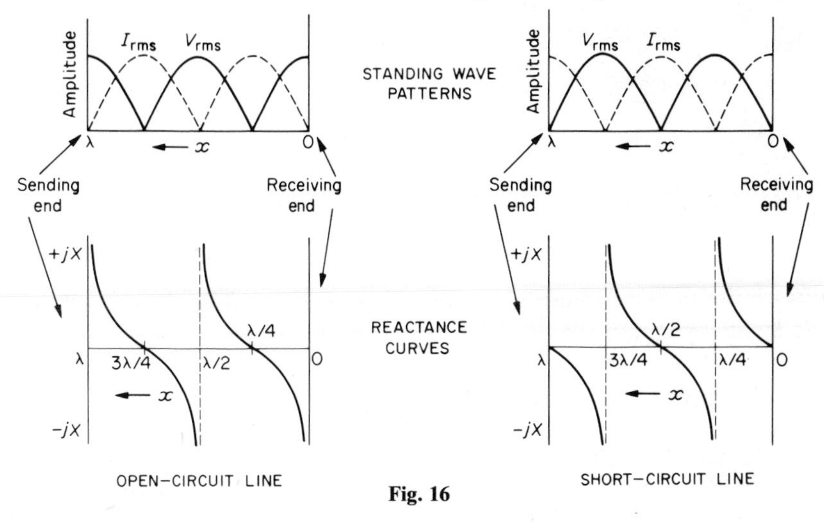

Fig. 16

seen that the input impedance of a $\lambda/4$ open-circuited line is zero, while that of a $\lambda/4$ short-circuited line is infinite.

3.3 Reflection coefficient

A reflection coefficient ρ at the receiving end is defined as the ratio of the reflected wave to the incident wave. At the end of a line of length l, the incident wave is $Ae^{-\gamma l}$ and the reflected wave $Be^{\gamma l}$. Hence

$$\rho = \frac{Be^{\gamma l}}{Ae^{-\gamma l}}$$

In general, ρ is complex and may be written as

$$\rho = |\rho| \, e^{j\phi}$$

where $|\rho|$ is its magnitude and ϕ its phase angle. It is related to the load impedance Z_R and the characteristic impedance Z_0 as may be seen from below. For the general line equations we have at $x = l$,

$$V_R = Ae^{-\gamma l} + Be^{\gamma l} = Ae^{-\gamma l} + \rho Ae^{-\gamma l}$$

$$I_R = A/Z_0 \,.\, e^{-\gamma l} - B/Z_0 \,.\, e^{\gamma l} = A/Z_0 \,.\, e^{-\gamma l} - \rho/Z_0 \,.\, Ae^{-\gamma l}$$

or

$$V_R = Ae^{-\gamma l}(1 + \rho)$$

$$Z_0 I_R = Ae^{-\gamma l}(1 - \rho)$$

or

$$Z_R/Z_0 = \frac{V_R}{I_R} = \frac{1 + \rho}{1 - \rho}$$

Hence

$$\rho = \frac{Z_R - Z_0}{Z_R + Z_0} \quad \text{where} \quad 0 \leqslant \rho \leqslant 1$$

Comments

1. For an open-circuited line, $Z_R = \infty$. Hence

$$\rho = \frac{1 - Z_0/Z_R}{1 + Z_0/Z_R} = 1$$

2. For a short-circuited line, $Z_R = 0$. Hence

$$\rho = \frac{Z_R - Z_0}{Z_R + Z_0} = \frac{-Z_0}{Z_0} = -1$$

and there is a phase reversal of voltage.

3. For a matched line, $Z_R = Z_0$. Hence

$$\rho = \frac{Z_0 - Z_0}{Z_0 + Z_0} = 0$$

and there is no reflected wave.

3.4 Voltage standing wave ratio (VSWR)

When a transmission line is terminated in an arbitrary impedance $Z_R \neq Z_0$, an incident and a reflected wave are both present on the line. If A and B are their respective amplitudes at some point on the line, a maximum occurs when the two amplitudes are in phase and a minimum, when the two amplitudes are 180° out of phase.

Hence

$$|V_{max}| = A + B = A(1 + B/A)$$
$$|V_{min}| = A - B = A(1 - B/A)$$

The standing wave ratio s is defined as the ratio of $|V_{max}|$ to $|V_{min}|$. Hence

$$s = \frac{|V_{max}|}{|V_{min}|} = \frac{1 + B/A}{1 - B/A}$$

Now the reflection factor at the receiving end was defined as

$$\rho = Be^{\gamma l}/Ae^{-\gamma l}$$

with

$$|\rho| = B/A$$

Substituting for B/A in the expression for s, we obtain

$$s = \frac{1 + |\rho|}{1 - |\rho|}$$

or

$$|\rho| = \frac{s - 1}{s + 1}$$

The standing wave ratio s is thus directly related to $|\rho|$ and since the latter is related to Z_R, we obtain

$$s = \frac{1 + |\rho|}{1 - |\rho|} = \left[1 + \left(\frac{Z_R - Z_0}{Z_R + Z_0}\right)\right] \bigg/ \left[1 - \left(\frac{Z_R - Z_0}{Z_R + Z_0}\right)\right] = \frac{Z_R}{Z_0}$$

or $\quad s = \dfrac{Z_R}{Z_0} \quad$ if $\quad Z_R > Z_0$

and $$s = \frac{Z_0}{Z_R} \quad \text{if} \quad Z_R < Z_0$$

Hence, measurements of s yield a direct knowledge of the nature of the load termination Z_R.

EXAMPLE 6

Describe the circumstances in which a standing wave can arise on a transmission line, and state the meaning of standing wave ratio.

An h.f. transmission line of negligible loss has a characteristic impedance of 600 Ω and is terminated by an antenna. Calculate the standing wave ratio along the line when the antenna impedance is

(a) 500 Ω

(b) 400 + j300 Ω (C & G Comm. Radio C, 1968)

Solution

The answer to the first part will be found in the text in Section 3.4.

(a) Since Z_R and Z_0 are pure resistances and $Z_R < Z_0$ we have

$$s = \frac{Z_0}{Z_R} = \frac{600}{500} = 1{\cdot}2$$

(b) $$\rho = \frac{Z_R - Z_0}{Z_R + Z_0} = \frac{400 + j300 - 600}{400 + j300 + 600} = \frac{-2 + j3}{10 + j3}$$

or $$|\rho| = \frac{\sqrt{2^2 + 3^2}}{\sqrt{10^2 + 3^2}} = 0{\cdot}345$$

Hence $$s = \frac{1 + |\rho|}{1 - |\rho|} = \frac{1{\cdot}345}{0{\cdot}654}$$

or $$s = 2{\cdot}06$$

3.5 Stub matching

When a line is 'matched', the reflection coefficient $\rho = 0$ and so the standing wave ratio $s = 1{\cdot}0$. Most systems are therefore designed to work with s as near to 1·0 as possible. A value of $s > 1$ represents mismatch and leads to loss of power at the receiving end. In other cases, it may cause a voltage breakdown as in high power radar systems or distortion, as in television. It is therefore important to be able to match a line.

Matching, in the case of a simple two-wire line, may be done by one or more stubs and is called 'stub-matching', or by the use of a quarter-wave transformer.* In the case of stub-matching, a stub is usually a piece of short-circuited line (assumed loss-less) which is placed in parallel with the line as near to the load as possible. Its position of attachment and length may be varied, as shown in Fig. 17.

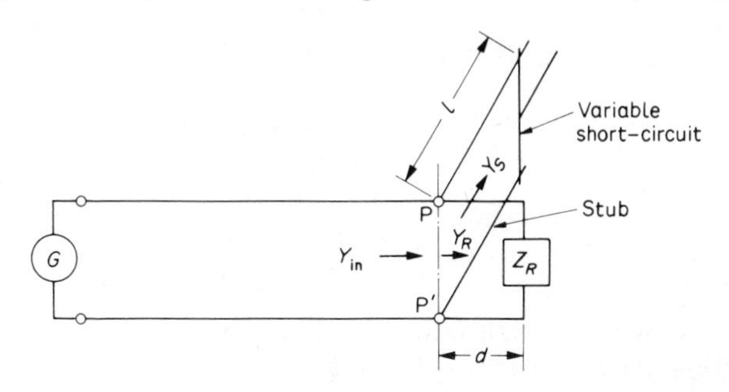

Fig. 17

By adjusting the position of the stub at PP′ and by varying the length of the short-circuited stub, standing waves can be eliminated left of PP′. If the input admittance at PP′ is Y_{in}, that of the load Y_R and that of the lossless stub Y_S, then $Y_{in} = Y_S + Y_R$

where
$$Y_R = \frac{1}{Z_R} = G_R \pm jB_R$$

and
$$Y_s = \frac{1}{Z_s} = \pm jB_s$$

The location PP′ is found by adjustment such that

$$G_R = G_0 = 1/Z_0$$

Hence
$$Y_{in} = G_0 \pm jB_R \pm jB_s$$

The length of the stub is varied until $|B_R| = |B_s|$ while their angles are opposite, for then the susceptancescancel, leaving a pure conductance G_0.

Hence

$$Y_{in} = G_0$$

or
$$Z_{in} = Z_0$$

* See *V.H.F. Line Techniques* by C. S. Gledhill, Edward Arnold (1960).

Usually the short-circuited stub is about $\lambda/4$ long since it can be made inductive or capacitive by varying its length near $l = \lambda/4$. The matching conditions may be obtained by calculation or with the aid of a Smith Chart as described hereafter.

(a) By calculation

The input impedance of a transmission line terminated in an impedance Z_R is

$$Z_{in} = Z_0\left[\frac{Z_R + Z_0 \tanh \gamma l}{Z_0 + Z_R \tanh \gamma l}\right]$$

or

$$Y_{in} = Y_0\left[\frac{Z_0 + Z_R \tanh \gamma l}{Z_R + Z_0 \tanh \gamma l}\right]$$

For a two-wire high-frequency line, $\tanh \gamma l = \mathrm{j} \tan \beta l$ and if the point of attachment is distant l_1 from the load then

$$Y_{in} = Y_0\left[\frac{Z_0 + \mathrm{j}Z_R \tan \beta l_1}{Z_R + \mathrm{j}Z_0 \tan \beta l_1}\right]$$

Rationalising then yields

$$Y_{in} = Y_0\left[\frac{(Z_0 + \mathrm{j}Z_R \tan \beta l_1)(Z_R - \mathrm{j}Z_0 \tan \beta l_1)}{(Z_R^2 + Z_0^2 \tan^2 \beta l_1)}\right]$$

$$= G_{in} + \mathrm{j}B_{in}$$

where

$$G_{in} = \frac{Z_R(1 + \tan^2 \beta l_1)}{(Z_R^2 + Z_0^2 \tan^2 \beta l_1)}$$

and

$$B_{in} = Y_0\left[\frac{(Z_R^2 - Z_0^2) \tan \beta l_1}{(Z_R^2 + Z_0^2 \tan^2 \beta l_1)}\right]$$

To match G_{in} we must have

$$G_{in} = \frac{1}{Z_0} = \frac{Z_R(1 + \tan^2 \beta l_1)}{(Z_R^2 + Z_0^2 \tan^2 \beta l_1)}$$

from which a solution for l_1 is obtained, since Z_0, Z_R and β are known.

Now, the input reactance X_s of a lossless short-circuited stub of length l_2 is given by

$$\mathrm{j}X_s = \mathrm{j}Z_0 \tan \beta l_2$$

or

$$\mathrm{j}B_s = -\mathrm{j}/(Z_0 \tan \beta l_2)$$

To obtain a match, the susceptances must cancel one another, i.e. they must be equal but of opposite sign.

Hence $$|B_{\text{in}}| = -|B_S|$$

or $$Y_0\left[\frac{(Z_R^2 - Z_0^2)\tan\beta l_1}{(Z_R^2 + Z_0^2\tan^2\beta l_1)}\right] = 1/(Z_0 \tan\beta l_2)$$

Hence $$\tan\beta l_2 = \left[\frac{Z_R^2 + Z_0^2\tan^2\beta l_1}{(Z_R^2 - Z_0^2)\tan\beta l_1}\right]$$

from which a value for l_2 can be obtained since Z_0, Z_R and l_1 are known.

(b) The Smith Chart[12]

It consists of two sets of circles, one of these is a polar coordinate system with its centre as the Chart centre and the other, is an orthogonal set of circles.

In the first set of circles, the radial distance represents the magnitude of the reflection factor ρ and the angular distance, its phase angle ϕ. The values of $|\rho|$ vary from 0 to 1·0 and ϕ from 0 to 180°. These coordinates are not normally drawn to avoid overcrowding the Chart and are shown dotted in Fig. 18 as an illustration. They also represent s and are called VSWR circles.

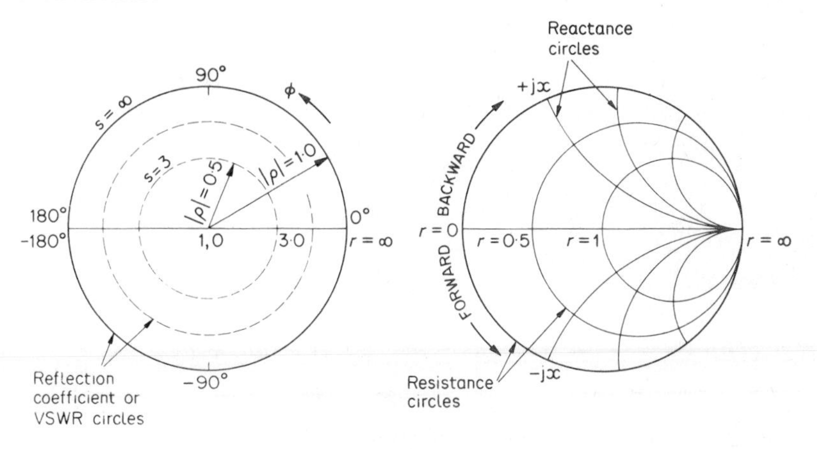

Fig. 18

The second set of circles shown in full lines in Fig. 18 which are orthogonal to one another, represent the resistive and reactive components of impedance whose values are all normalised. The normalised resistance r and reactance x are given by

$$r = R/Z_0$$

$$x = X/Z_0$$

where R, X are the actual values and r, x are the normalised values obtained by dividing the actual values by Z_0, the characteristic impedance. Both r and x range from 0 to ∞, the outermost circle of the Chart corresponding to $r = 0$. The centres of these circles lie on two perpendicular lines and all these circles pass through the point ∞.

Circles of resistance are drawn in fully, while only part of the circles of reactance appear on the Chart. The intersection of a resistance circle and a reactance circle gives a point of normalised impedance z and its inversion about the centre gives the normalised admittance y.

Hence
$$z = \frac{Z}{Z_0} = r \pm jx$$

or
$$y = \frac{Y}{Y_0} = g \pm jb$$

The construction of these orthogonal circles can be obtained by a consideration of any load impedance Z_R. The normalised impedance is given by

$$z = \frac{Z_R}{Z_0} = s = \frac{1 + \rho}{1 - \rho} = r + jx$$

Since ρ is complex we also have

$$\rho = a + jb$$

and
$$z = \frac{1 + a + jb}{1 - a - jb} = r + jx$$

Hence

$$r + jx = \frac{(1 + a) + jb}{(1 - a) - jb} = \frac{(1 - a^2 - b^2) + 2jb}{(1 - a)^2 + b^2}$$

or
$$r = \frac{1 - a^2 - b^2}{(1 - a)^2 + b^2}$$

$$x = \frac{2b}{(1 - a)^2 + b^2}$$

Completing the squares yields

$$\left[a - \frac{r}{(r+1)}\right]^2 + b^2 = \frac{1}{(r+1)^2} \tag{1}$$

and

$$(a-1)^2 + \left[b - \frac{1}{x}\right]^2 = \frac{1}{x^2} \tag{2}$$

Equation (1) represents a set of resistance circles with centre $[r/(r+1), 0]$, radius $1/(r+1)$ and are drawn in Fig. 19, while the last equation (2) represents a set of reactance circles with centre $[1, 1/x]$, radius $1/x$ and are also shown in Fig. 19.

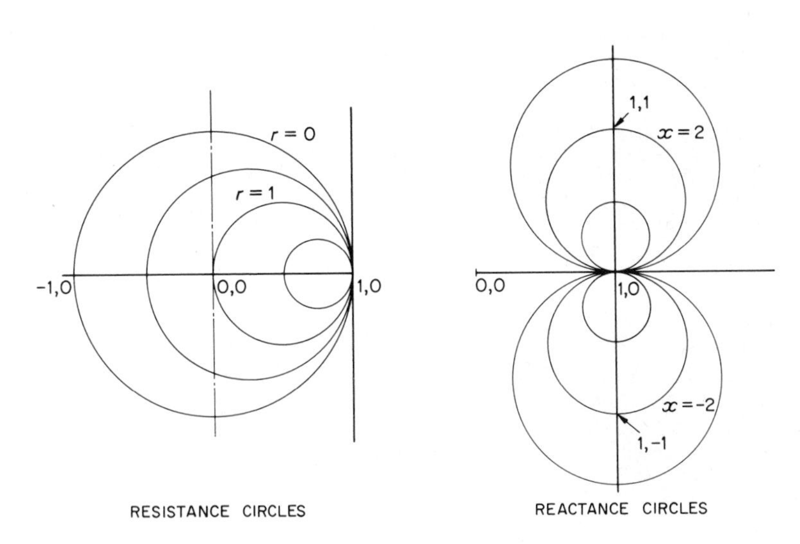

Fig. 19

The final Smith Chart is drawn by moving the x co-ordinate (by unity) and using the relative Chart centre (1, 0) rather than (0, 0), with the reactance circles drawn in only partially as stated earlier. This is illustrated in Fig. 20.

The outermost circle is scaled off in wavelengths 0 to $0{\cdot}5\lambda$ towards the generator (backwards) or 0 to $0{\cdot}5\lambda$ towards the load (forwards). Only half a wavelength is scaled off since the standing wave pattern repeats itself every $\lambda/2$ as illustrated in Fig. 21.

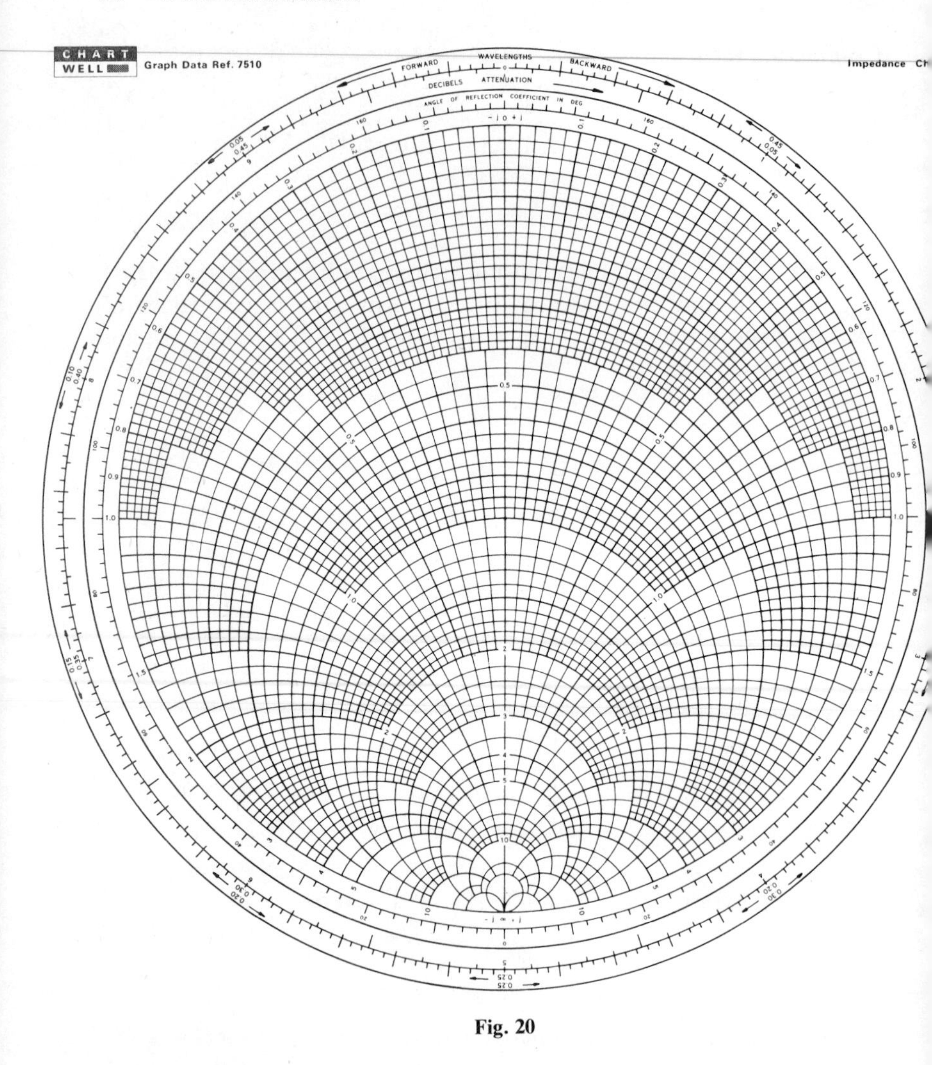

Fig. 20

(c) Typical examples

Load impedance

To plot an impedance 50 + j100 Ω on a 50 Ω line.

1. Normalising Z_R yields z_R = (50 + j100)/50 = 1 + j2·0 Ω.
2. Set chart with the central diameter horizontal and the point zero on

the left. Starting from $r = 0$ on the central diameter move to point $r = 1{\cdot}0$ on the right.
3. Follow resistance circle through point 1·0 upward.
4. Locate point where it crosses reactance circle j2·0. Point of intersection is $z = 1{\cdot}0 + j2{\cdot}0\ \Omega$.

Voltage standing wave ratio

Let the load impedance be $100 - j50\ \Omega$ on a $50\ \Omega$ line. To find the VSWR.
1. Normalising Z_R yields $z_R = (100 - j50)/50 = 2 - j1{\cdot}0\ \Omega$.
2. Plot it on the Chart.
3. Draw a circle with centre at point (1, 0) through point z_R.
4. Read point of intersection of the circle with the horizontal diameter on the right of the Chart centre. This gives the VSWR as 2·6.

Reflection coefficient

Let the load impedance be $100 + j75\ \Omega$ on a $50\ \Omega$ line. To find the reflection coefficient.
1. Normalising Z_R yields $z_R = (100 + j75)/50 = 2 + j1{\cdot}5\ \Omega$.
2. Plot z_R.
3. Draw the VSWR circle through z_R and read the VSWR as 3·3.
4. From $|\rho| = (s - 1)/(s + 1)$, $|\rho|$ is obtained as 0·535.
5. Alternatively, scale-off the radius of the VSWR circle through point '3·3' and the unit circle radius also. The ratio of the smaller to larger radius is $|\rho| = 0{\cdot}535$.

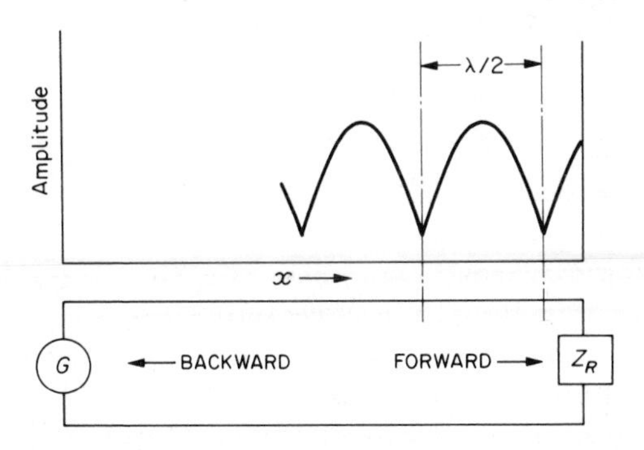

Fig. 21

6. Draw a radial line from Chart centre through z_R to meet the phase angle circle. Read ϕ as 30°. Hence $\rho = 0{\cdot}535\underline{/30°}$.

Admittance of a load

Let the load be $Z_R = 150 - j75\ \Omega$ on a $50\ \Omega$ line. To find Y_R.

1. Normalising Z_R yields $z_R = (150 - j75)/50 = 3 - j1{\cdot}5\ \Omega$.
2. Plot z_R.
3. Draw the VSWR circle through z_R.
4. From point z_R draw a diameter through the Chart centre to intersect the circle again.
5. Point of intersection is $y_R = 0{\cdot}27 + j0{\cdot}14$ which is the required normalised admittance.
6. Admittance $Y_R = y_R/50 = 0{\cdot}0054 + j0{\cdot}0028S$.

Distance from load to voltage minimum

Let the load be $Z_R = 50 + j150\ \Omega$ on a $50\ \Omega$ line. To find the distance from Z_R to the first voltage minimum.

1. Normalising Z_R yields $z_R = (50 + j150)/50 = 1 + j3{\cdot}0\ \Omega$.
2. Plot z_R.
3. Draw radial line from Chart centre through z_R to intersect wavelengths circle.
4. Locate point of voltage minimum as point $r = 0$ on the left of the Chart. Note position on wavelengths circle as zero λ.
5. Move along wavelengths circle 'forwards'.
6. Read wavelength at point of intersection with radial line through z_R as $0{\cdot}296\lambda$. Hence, the first voltage minimum is $0{\cdot}296\lambda$ from the load.

Stub matching

To determine the location and length of a single stub to match a $50\ \Omega$ line terminated in $Z_R = 100 - j50\ \Omega$.

1. Normalising Z_R yields $z_R = (100 - j50)/50 = 2 - j1{\cdot}0\ \Omega$.
2. Plot z_R and draw the VSWR circle through z_R.
3. Determine admittance $y_R = 0{\cdot}4 + j0{\cdot}2$, and draw the radial line through y_R to intersect wavelengths circle.
4. Locate its position relative to arbitrary zero λ by reading wavelengths 'backwards' as $0{\cdot}037\lambda$.

5. Move around VSWR circle to meet intersection with resistance circle $r = 1{\cdot}0$. Point of intersection is $y_R = 1{\cdot}0 + j1{\cdot}0S$.
6. Determine its position relative to arbitrary zero λ by reading 'backwards' a value as $0{\cdot}162\lambda$.
7. Location of stub from load equals $0{\cdot}162\lambda - 0{\cdot}037\lambda = 0{\cdot}125\lambda$.
8. To determine length of stub we require to cancel $j1{\cdot}0S$ with $-j1{\cdot}0S$. Locate point $-j1{\cdot}0S$ on VSWR $= \infty$ circle.
9. Draw radial line through it to intersect wavelengths circle.
10. Input admittance *at* a short-circuit is $j\infty$. Locate point $j\infty$ ($0{\cdot}25\lambda$ position) at right of Chart and travel along wavelengths circle to meet radial line through $-j1{\cdot}0S$.
11. Read 'backwards' on wavelengths circle the value as $0{\cdot}375\lambda$.
12. Length of short-circuited stub is $0{\cdot}375\lambda - 0{\cdot}25\lambda = 0{\cdot}125\lambda$.

Load impedance from shift of minima

To find the load impedance when a voltage minimum shifts $0{\cdot}1\lambda$ toward the load when the load is shorted. Let the VSWR on the 50 Ω line be 2·0.
1. When the load is shorted, first minimum occurs at the load. Since it moves $0{\cdot}1\lambda$ towards the load when the load is shorted, it must have been $0{\cdot}1\lambda$ away from the load.
2. Locate the short circuit $r = 0$ at the left hand end on the central Chart diameter.
3. Move $0{\cdot}1\lambda$ towards load 'forwards', along wavelengths circle and draw radial line to Chart centre.
4. Draw a VSWR circle of 2·0. Point of intersection with radial line is the load impedance $z_R = 0{\cdot}68 - j0{\cdot}48\ \Omega$.
5. Impedance $Z_R = 50(0{\cdot}68 - j0{\cdot}48) = 34{\cdot}0 - j24\ \Omega$.

EXAMPLE 7
The voltage and current for a travelling wave on a transmission line of propagation coefficient γ and characteristic impedance Z_0 are $V(x) = A\exp(-\gamma x)$ and $I(x) = A\exp(-\gamma x)/Z_0$ respectively, A being an arbitrary constant.

The line is terminated by a load impedance ζZ_0. Show that the impedance at distance l from the load is

$$Z_{in} = \frac{Z_0(\zeta + \tanh \gamma l)}{1 + \zeta \tanh \gamma l}$$

Outline the basic principles from which the Smith Chart is developed and indicate how this Chart is used in impedance-matching problems.

A loss-less 50 Ω transmission line is terminated by an unknown impedance. The first voltage minimum is at a distance $0{\cdot}2\lambda$ from the termination and the VSWR on the line is 2. Determine the impedance.

(C.E.I., Pt. 2, Comm. Eng. 1970)

Solution

The first part of the solution will be found in Section 2.6 which may be converted to the form given in the problem by the substitution of $Z_{in} = Z_S$ and $\zeta = Z_R/Z_0$.

The second part of the solution has been considered in Sections 3.5(b) and 3.5(c).

Final part

Since a voltage minimum is $0{\cdot}2\lambda$ from the termination, locate on the Chart the point $r = 0$ on the left of the central diameter and proceed $0{\cdot}2\lambda$ 'forwards' along the wavelengths circle. Draw a radial line from the Chart centre to this point.

Next, draw the VSWR circle of 2 and the point of intersection of this circle with the previous radial line is z_R. Read z_R as $1{\cdot}6 - j0{\cdot}7$. Hence $Z_R = 50(1{\cdot}6 - j0{\cdot}7) = 80 - j35$ Ω.

EXAMPLE 8

A 600 Ω, loss-free unit-phase velocity transmission line 105 ft long is used to connect a 200 MHz transmitter, having an output impedance of $600\angle 0^\circ$ to an antenna with a terminal impedance of $700\angle -72^\circ$. Using either a Smith Chart or other means, determine

(a) the VSWR on the antenna feeder.
(b) the voltage reflection coefficient at the antenna terminals.
(c) the impedance presented at the transmitter terminals.
(d) the distance from the transmitter terminals at which a matching stub could be connected.
(e) the reactance of the stub.

(C. & G. Adv. Comm. Radio, 1970)

Solution

$$f\lambda = 3 \times 10^8$$

or
$$\lambda = \frac{3 \times 10^8}{2 \times 10^8} = 1{\cdot}5 \text{ m}$$

$$\text{Line length} = \frac{105}{1{\cdot}5 \times 3{\cdot}27} = 21{\cdot}4\lambda$$

Normalised load impedance

$$z_R = \tfrac{700}{600}[\cos(-72) + \text{j} \sin(-72)]$$
$$= \tfrac{7}{6} \times (0{\cdot}309 - \text{j}0{\cdot}9511)$$

or
$$z_R = 0{\cdot}36 - \text{j}1{\cdot}11\ \Omega$$

Procedure

1. Plot z_R on the Smith Chart.
2. Draw VSWR circle through z_R and read VSWR as 6·4.
3. $|\rho| = (s - 1)/(s + 1) = 5{\cdot}4/7{\cdot}4 = 0{\cdot}73$.
4. Impedance z_R is $0{\cdot}363\lambda$ 'backwards'.
5. Input impedance is $21{\cdot}4\lambda$ from load which is the same as being $0{\cdot}4\lambda$ from the load. Hence input Z is $0{\cdot}4\lambda$ from Z_R towards generator (backwards) which yields $z_{in} = 5{\cdot}1 - \text{j}2{\cdot}5\ \Omega$. Hence, impedance from transmitter terminals $Z = 600(5{\cdot}1 - \text{j}2{\cdot}5) = 3060 - \text{j}1500\ \Omega$.
6. Load admittance $y_R = 0{\cdot}28 + \text{j}0{\cdot}82S$.
7. Radial line through y_R yields $0{\cdot}113\lambda$ towards generator.
8. To match y_R, move around VSWR circle to intersect the circle through point (1, 0). Read new admittance y'_R as $1 + \text{j}2{\cdot}1$. Hence we require a stub susceptance of $-\text{j}2{\cdot}1$ to neutralise the line susceptance of $+\text{j}2{\cdot}1$. Hence, stub *reactance* is $1/-\text{j}2{\cdot}1 = \text{j}0{\cdot}476\ \Omega$.
9. Draw a radial line through y'_R to give a value of $0{\cdot}19\lambda$ 'backwards' on wavelengths circle.
10. Position of stub location is $(0{\cdot}19 - 0{\cdot}113)\lambda = 0{\cdot}077\lambda$ from load. Distance of stub from transmitter terminals is $21{\cdot}4\lambda - 0{\cdot}077\lambda = 21{\cdot}32\lambda$.

4
Field phenomena

4.1 Electromagnetic waves

Electromagnetic waves may be transmitted along some form of guided structure and are called guided waves. Alternatively, they may be propagated through free space and are therefore unguided waves. In both cases, energy is carried by the electric and magnetic fields associated with the wave, hence the name electromagnetic wave.

However, at lower power frequencies it is more usual to talk about voltages and currents along a transmission line rather than the fields of the wave. Nevertheless, the energy is associated with the fields and the transmission line merely serves to guide the energy along. Such is the case for the two-wire line and coaxial cable, and the form of propagation is called a Transverse Electric and Magnetic wave or TEM wave as briefly outlined in the first chapter.

In this type of propagation the presence of the field is depicted by field lines whose direction is indicated by arrows and its intensity by the density of lines drawn. Typical field configurations for a two-wire line and a coaxial cable are shown in Fig. 22.

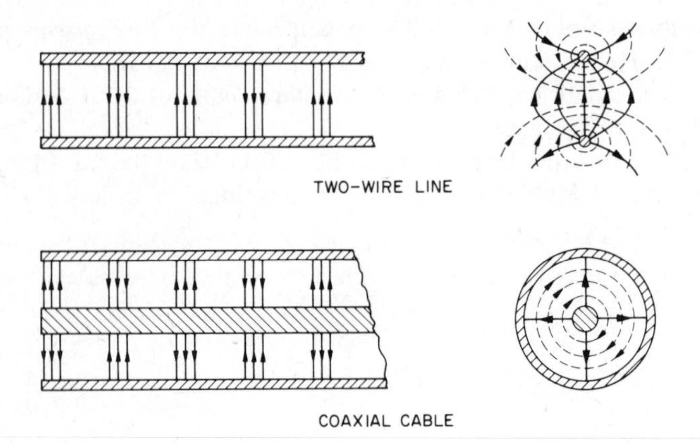

Fig. 22

At very high frequencies around 1 GHz, electromagnetic waves may under certain circumstances be transmitted along hollow metal guides called waveguides. Such guided waves are usually associated with fields rather than voltages and currents and the form of field propagation, though guided, is somewhat different from that of a TEM wave. It is called either a transverse electric (TE or H wave) or transverse magnetic (TM or E wave) as illustrated in Fig. 23.

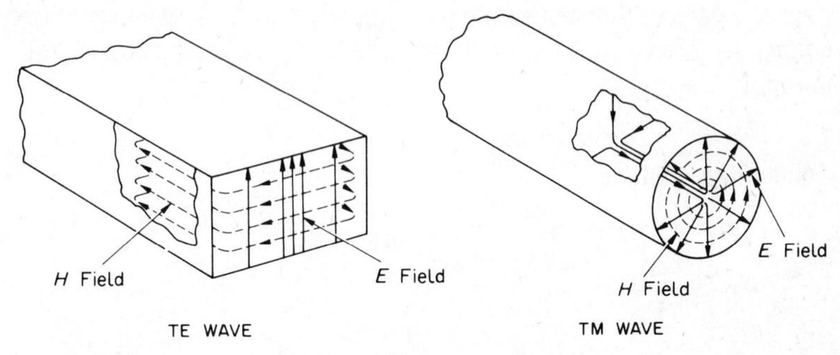

Fig. 23

Nevertheless, the basis of all such field phenomena is described in terms of certain vector quantities **E, D, B, H, J** which are formulated in a general form of field theory. The basis of such field theory may be purely electric such as the well known theory of Electrostatics or it may be purely magnetic and is known as Magnetostatics. A combination of both electric and magnetic effects due to moving charges or currents is called Electromagnetic Field Theory. Details of important concepts of Electrostatic and Magnetostatic theories are given by Ramo.[13] The development of electromagnetic field theory will now be considered.

4.2 Electromagnetic field theory

It is based on the five field vectors **E, D, B, H, J,** whose basic equations are given by Ramo.[13] It is usual to refer to these quantities as vectors since they have both magnitude and direction. The essential study concerning the behaviour of vectors is known as vector analysis and is detailed in Appendix A. The basic ideas of electrostatic theory and magnetostatic theory were taken further and combined into a single unified theory of electromagnetic waves by Maxwell and are concisely expressed in his well known equations.

(a) Maxwell's equations

These are four basic equations which apply to time-varying (alternating) fields and electromagnetic phenomena. The theory is a generalisation of the previous work of Faraday, Ampere and Gauss. Maxwell was able to build upon their ideas and to make a further contribution himself which led him to formulating a coherent theory for the field vectors **E, D, B, H** and **J.** Further details are given by Stratton.[14]

In the case of alternating fields **E** and **H,** we can summarise Maxwell's equations for a conducting medium and a dielectric medium in the following manner.

Conducting medium

$$\text{curl } \mathbf{H} = \mathbf{J} + \frac{\partial \mathbf{D}}{\partial t} = (\sigma + \mathrm{j}\omega\varepsilon)\mathbf{E}$$

$$\text{curl } \mathbf{E} = -\frac{\partial \mathbf{B}}{\partial t} = -\mathrm{j}\omega\mu\mathbf{H}$$

$$\text{div } \mathbf{D} = \rho$$

$$\text{div } \mathbf{B} = 0$$

Dielectric medium

$$\text{curl } \mathbf{H} = \frac{\partial \mathbf{D}}{\partial t} = \mathrm{j}\omega\varepsilon\mathbf{E}$$

$$\text{curl } \mathbf{E} = \frac{\partial \mathbf{B}}{\partial t} = -\mathrm{j}\omega\mu\mathbf{H}$$

$$\text{div } \mathbf{D} = 0$$

$$\text{div } \mathbf{B} = 0$$

since σ and ρ are both zero in a dielectric medium.

(b) The wave equation

For a dielectric medium we have

$$\text{curl } \mathbf{E} = \nabla \times \mathbf{E} = -\mathrm{j}\omega\mu\mathbf{H}$$

Hence $\nabla \times \nabla \times \mathbf{E} = \text{curl}(-j\omega\mu\mathbf{H}) = -j\omega\mu \text{ curl } \mathbf{H}$

or $\nabla \times \nabla \times \mathbf{E} = \omega^2\mu\varepsilon\mathbf{E}$

Also $\nabla(\nabla \cdot \mathbf{E}) - \nabla^2\mathbf{E} = \nabla \times \nabla \times \mathbf{E}$

with $\nabla \cdot \mathbf{D} = \nabla \cdot \varepsilon\mathbf{E} = 0$

Hence $\nabla(0) - \nabla^2\mathbf{E} = \omega^2\mu\varepsilon\mathbf{E}$

or $\nabla^2\mathbf{E} = -\omega^2\mu\varepsilon\mathbf{E}$

Now $1/\sqrt{\mu\varepsilon}$ has the dimensions of a velocity from which Maxwell concluded that light is an electromagnetic wave propagated through free space with velocity $c = 1/\sqrt{\mu\varepsilon}$.

Hence $\nabla^2\mathbf{E} = (-\omega^2/c^2)\mathbf{E}$

For alternating fields we have also

$$\mathbf{E} = \mathbf{E}_0 \sin \omega t$$

or $$\frac{\partial^2\mathbf{E}}{\partial t^2} = -\omega^2\mathbf{E}_0 \sin \omega t = -\omega^2\mathbf{E}$$

Hence $$\nabla^2\mathbf{E} = -1/c^2 \frac{\partial^2\mathbf{E}}{\partial t^2}$$

Similarly $$\nabla^2\mathbf{H} = -\frac{1}{c^2}\frac{\partial^2\mathbf{H}}{\partial t^2}$$

which are the wave equations for the alternating field vectors **E** and **H**.

In a *conducting* medium without free electrical charges, we have finite σ with $\rho = 0$.

Hence $\nabla \times \mathbf{E} = -j\omega\mu\mathbf{H}$

or $\nabla \times \nabla \times \mathbf{E} = -j\omega\mu \text{ curl } \mathbf{H}$

$= -j\omega\mu(\sigma + j\omega\varepsilon)\mathbf{E}$

Since $\nabla \times \nabla \times \mathbf{E} = \nabla(\nabla . \mathbf{E}) - \nabla^2\mathbf{E}$

with $\nabla . \mathbf{D} = \nabla . \varepsilon\mathbf{E} = 0$

Hence $\nabla^2\mathbf{E} = -\nabla \times \nabla \times \mathbf{E} = (j\omega\mu\sigma - \omega^2\mu\varepsilon)\mathbf{E}$

or $\nabla^2\mathbf{E} = \gamma^2\mathbf{E}$

where $\gamma^2 = (j\beta)^2 = -\beta^2 = (j\omega\mu\sigma - \omega^2\mu\varepsilon)$

or $\beta = \sqrt{\omega^2\mu\varepsilon - j\omega\mu\sigma}$

Similarly $\nabla^2 \mathbf{H} = \gamma^2 \mathbf{H}$

Also since $\mathbf{E} = \mathbf{E}_0 \sin \omega t$

$$\frac{\partial \mathbf{E}}{\partial t} = \omega \mathbf{E}_0 \cos \omega t = \mathrm{j}\omega \mathbf{E}$$

$$\frac{\partial^2 \mathbf{E}}{\partial t^2} = -\omega^2 \mathbf{E}_0 \sin \omega t = -\omega^2 \mathbf{E}$$

with $\omega^2 \mu \varepsilon = \omega^2/c^2 = -1/c^2 \dfrac{\partial^2}{\partial t^2}$

Hence
$$\nabla^2 \mathbf{E} = \gamma^2 \mathbf{E} = \sigma\mu \frac{\partial \mathbf{E}}{\partial t} + \frac{1}{c^2}\frac{\partial^2 \mathbf{E}}{\partial t^2}$$

$$\nabla^2 \mathbf{H} = \gamma^2 \mathbf{H} = \sigma\mu \frac{\partial \mathbf{H}}{\partial t} + \frac{1}{c^2}\frac{\partial^2 \mathbf{H}}{\partial t^2}$$

(c) Power flow

The energy stored per unit volume in an electrostatic field is $\frac{1}{2}\varepsilon \mathbf{E}^2$ where **E** is the electric field vector. For a magnetic field it is $\frac{1}{2}\mu \mathbf{H}^2$ where **H** is the magnetic field vector. For an electromagnetic field, the most general expression is the sum of the electric and magnetic energies.

Hence
$$\left.\begin{matrix}\text{Total energy stored}\\ \text{per unit volume}\end{matrix}\right\} = \tfrac{1}{2}(\varepsilon \mathbf{E}^2 + \mu \mathbf{H}^2)/m^3$$

Consider now, the surface **da** shown in Fig. 24 with outward flow $\mathscr{P}$ per unit area.

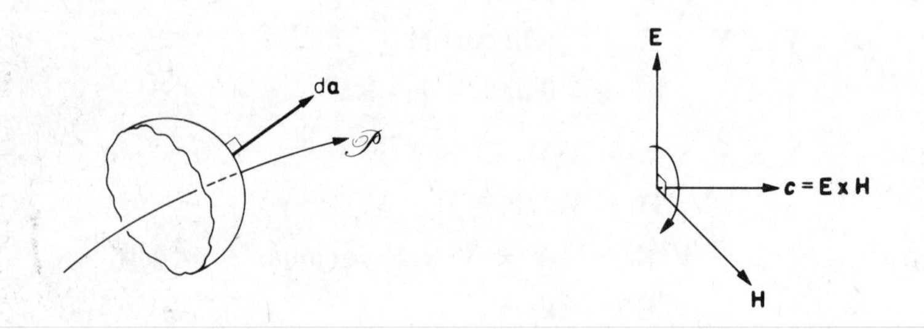

Fig. 24

Total power flow through d**a** $= \oint_S \mathscr{P} \cdot d\mathbf{a}$. This leads to a loss of stored energy given by

$$\text{Loss} = \frac{-\partial}{\partial t}\left[\tfrac{1}{2}\oint_V (\varepsilon E^2 + \mu H^2)\, dv\right] \quad \text{(magnitude only)}$$

or
$$\oint_S \mathscr{P} \cdot d\mathbf{a} = -\oint_V \left(\mu H \frac{\partial H}{\partial t} + \varepsilon E \frac{\partial E}{\partial t}\right) dv$$

$$= -\oint_V \left(\mu \mathbf{H} \cdot \frac{\partial \mathbf{H}}{\partial t} + \varepsilon \mathbf{E} \cdot \frac{\partial \mathbf{E}}{\partial t}\right) dv$$

But
$$\nabla \times \mathbf{E} = -\mu \frac{\partial \mathbf{H}}{\partial t} \quad \text{and} \quad \nabla \times \mathbf{H} = \varepsilon \frac{\partial \mathbf{E}}{\partial t}$$

for a dielectric medium.

Hence
$$\oint_S \mathscr{P} \cdot d\mathbf{a} = \oint_V [\mathbf{H} \cdot (\nabla \times \mathbf{E}) - \mathbf{E} \cdot (\nabla \times \mathbf{H})]\, dv$$

Also
$$\nabla \cdot (\mathbf{E} \times \mathbf{H}) = \mathbf{H} \cdot (\nabla \times \mathbf{E}) - \mathbf{E} \cdot (\nabla \times \mathbf{H})$$

Hence
$$\oint_S \mathscr{P} \cdot d\mathbf{a} = \oint_V \nabla \cdot (\mathbf{E} \times \mathbf{H})\, dv = \oint_S (\mathbf{E} \times \mathbf{H})\, d\mathbf{a}$$

or
$$\mathscr{P} = \mathbf{E} \times \mathbf{H} \text{ watts/m}^2$$

which is Poynting's theorem.

For alternating quantities we obtain

$$\mathscr{P}_{\text{av}} = \tfrac{1}{2}(\mathbf{E} \times \mathbf{H}^*) \text{ watts/m}^2$$

where $\mathbf{H}^*$ is the conjugate of **H**. This gives the *real power* propagated with velocity **c** as shown in Fig. 24.

EXAMPLE 9

A plane TEM wave is travelling through free space in the X direction with velocity c. Assuming sinusoidal variations of the electric and magnetic fields, deduce the intrinsic impedance of free space.

Solution

Assuming rectangular coordinates, for the **E** vector we have

$$\mathbf{E} = \mathbf{i}E_x + \mathbf{j}E_y + \mathbf{k}E_z$$

with $$\nabla^2 \mathbf{E} = \frac{1}{c^2}\frac{\partial^2 \mathbf{E}}{\partial t^2}$$

Hence $$\frac{\partial^2 E_x}{\partial x^2} + \frac{\partial^2 E_x}{\partial y^2} + \frac{\partial^2 E_x}{\partial z^2} = \frac{1}{c^2}\frac{\partial^2 E_x}{\partial t^2}$$

$$\frac{\partial^2 E_y}{\partial x^2} + \frac{\partial^2 E_y}{\partial y^2} + \frac{\partial^2 E_y}{\partial z^2} = \frac{1}{c^2}\frac{\partial^2 E_y}{\partial t^2}$$

$$\frac{\partial^2 E_z}{\partial x^2} + \frac{\partial^2 E_z}{\partial y^2} + \frac{\partial^2 E_z}{\partial z^2} = \frac{1}{c^2}\frac{\partial^2 E_z}{\partial t^2}$$

Since the plane wave is travelling along the X direction, it has only components $\mathbf{E}_y$ and $\mathbf{H}_z$ as shown in Fig. 25.

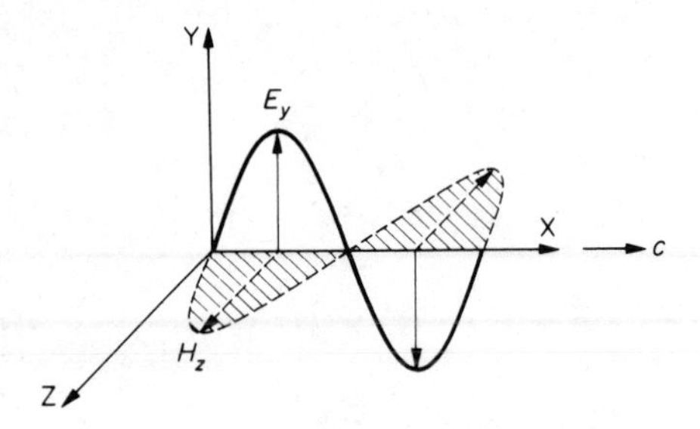

Fig. 25

Hence $$E_x = E_z = 0$$

$$\frac{\partial}{\partial y} = \frac{\partial}{\partial z} = 0$$

with $$\frac{\partial^2 E_y}{\partial x^2} = \frac{1}{c^2}\frac{\partial^2 E_y}{\partial t^2}$$

This last equation is similar to that for a transmission line. The general solution is of form

$$E_y = (A\mathrm{e}^{-\gamma x} + B\mathrm{e}^{\gamma x})\mathrm{e}^{\mathrm{j}\omega t}$$

As there is no reflected wave in free space, we obtain

$$E_y = A\mathrm{e}^{-\gamma x}\,\mathrm{e}^{\mathrm{j}\omega t}$$

If the peak value of the field is E_0 at $x = 0$, then

$$E_y = E_0\mathrm{e}^{-\gamma x}\,\mathrm{e}^{\mathrm{j}\omega t}$$

For free space $\alpha \simeq 0$ and so $\gamma = \mathrm{j}\beta$ with $\beta = 2\pi/\lambda$ where λ is the free space wavelength.

Hence $$E_y = E_0 \mathrm{e}^{-\mathrm{j}\beta x}\,\mathrm{e}^{\mathrm{j}\omega t} = E_0 \mathrm{e}^{\mathrm{j}(\omega t - \beta x)}$$

Similarly, starting from the equation

$$\frac{\partial^2 H_z}{\partial x^2} = \frac{1}{c^2}\cdot\frac{\partial^2 H_z}{\partial t^2}$$

we obtain $H_z = H_0 \mathrm{e}^{\mathrm{j}(\omega t - \beta x)}$

H_z component

Since curl $\mathbf{H} = \varepsilon_0(\partial \mathbf{E}/\partial t)$ for free space, we have

$$-\mathrm{j}\frac{\partial H_z}{\partial x} = \mathrm{j}\varepsilon_0(\partial/\partial t)[E_0 \mathrm{e}^{\mathrm{j}(\omega t - \beta x)}]$$

$$= \mathrm{j}\varepsilon_0 \mathrm{j}\omega E_0 \mathrm{e}^{\mathrm{j}(\omega t - \beta x)}$$

$$= \mathrm{j}\varepsilon_0 \mathrm{j}\omega E_y$$

or $$\frac{\partial H_z}{\partial x} = -\varepsilon_0 \mathrm{j}\omega E_y$$

so $$-\mathrm{j}\beta H_0 \mathrm{e}^{-\mathrm{j}\beta x}\,\mathrm{e}^{\mathrm{j}\omega t} = -\mathrm{j}\beta H_z = -\varepsilon_0 \mathrm{j}\omega E_v$$

with $$H_z = \frac{-\varepsilon_0 \mathrm{j}\omega E_y}{-\mathrm{j}\beta} = \varepsilon_0(\omega/\beta)E_y = \varepsilon_0 c E_y$$

Hence $$\frac{E_y}{H_z} = \frac{E_y}{\varepsilon_0 c E_y} = \frac{1}{\varepsilon_0 c} = \sqrt{\frac{\mu_0}{\varepsilon_0}} = 377\,\Omega$$

or $$Z_0 = 377\,\Omega$$

where Z_0 is called the intrinsic impedance of free space.

5
Waveguide theory

Electromagnetic waves may be transmitted along hollow tubes or waveguides under certain conditions. Such guided waves are unlike those transmitted by two-wire lines or coaxial cables and are of main interest at microwave frequencies.

The waveguide is essentially a hollow metal tube either of rectangular or circular cross section as shown in Fig. 23 and is made of copper, brass or aluminium. For special applications, other shapes of guide may also be used, e.g. square waveguides. The mode of propagation is essentially in terms of the electric and magnetic field as there is no centre conductor, while the hollow guide merely serves to guide the electromagnetic wave. Occasionally, reference may be made to the voltage across the guide or to the induced currents in the walls of the guide as they throw some light on power transmitted by the waveguide or on the power losses in the walls of the waveguide.

5.1 Waveguide transmission[15,16]

It is characterised by two important features:

(a) There is a minimum frequency below which a given waveguide will not transmit the wave. It is called the cut-off frequency f_c and is directly related to the size of waveguide used.

(b) There is always a component of **E** or **H** along the direction of propagation.

The guided waves may be propagated along the waveguide with different field patterns called 'modes'. Modes are mainly of two types, those which have an **E** component along the direction of propagation are called TM or *E waves*, while those with an **H** component along the direction of propagation are called TE or *H waves*. To distinguish between the various types of modes, subscripts m and n are used as defined hereafter.

Rectangular modes

The two types are $TE_{mn}(H_{mn})$ modes and $TM_{mn}(E_{mn})$ modes, where m, n are integers. The subscript m refers to the number of *half* sinusoidal variations of the field along dimension 'a', while the subscript n refers to the number of *half* sinusoidal variations of the field along dimension 'b', e.g. $TE_{10}(H_{10})$, $TM_{11}(E_{11})$.

Circular modes

The two types are $TE_{mn}(H_{mn})$ modes and $TM_{mn}(E_{mn})$ modes where m, n are integers. The subscript m refers to the number of *full* sinusoidal variations of the field along the circumference and the subscript n refers to the number of *half* sinusoidal variations of the field along a radius, e.g. $TE_{01}(H_{01})$, $TM_{12}(E_{12})$.

Basically, a rectangular waveguide may be considered as a form of parallel plate transmission line between which the wave is trapped. It is therefore possible to propagate a waveguide mode by the synthesis of two plane waves travelling towards each other at an angle θ to the horizontal provided the boundary conditions given in Appendix B are satisfied.

Consider for example the two plane waves shown in Fig. 26 which are travelling at an angle θ with the horizontal. For constructional reasons, magnetic field lines are considered, spaced $\lambda/2$ apart rather than electric field lines.

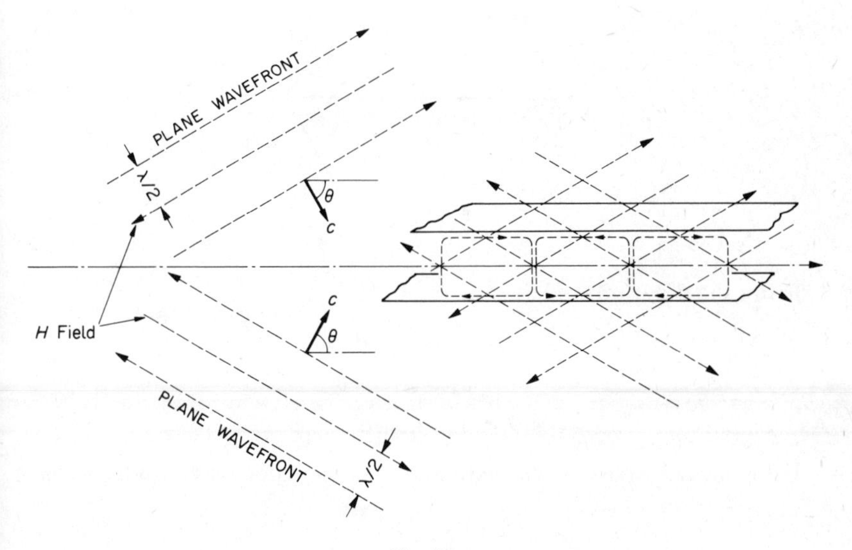

Fig. 26

When the two waves overlap as shown on the right, the resultant magnetic field lines at the intersecting points form closed loops along the axis of the diagram. By placing parallel plates tangential to the sides of a single set of such loops, a waveguide mode can be shown to exist within such a parallel plate structure. Side plates may then be added to form a closed waveguide with the wave travelling from left to right along the waveguide. The predominant $TE_{10}(H_{10})$ mode is shown in Fig. 29.

5.2 Phase and group velocities

The component plane waves travel with the velocity of light c towards one another, but the composite wave pattern formed when they overlap, travels along the guide with a different velocity called the phase or guide velocity v_p.

To illustrate this, let the wavefronts be represented by OM and ON as in Fig. 27 and after one second let them travel through distances MP and NP at angle θ to the axis. In the same time, the point of intersection 0 on the wavefronts has travelled to P with velocity v_p.

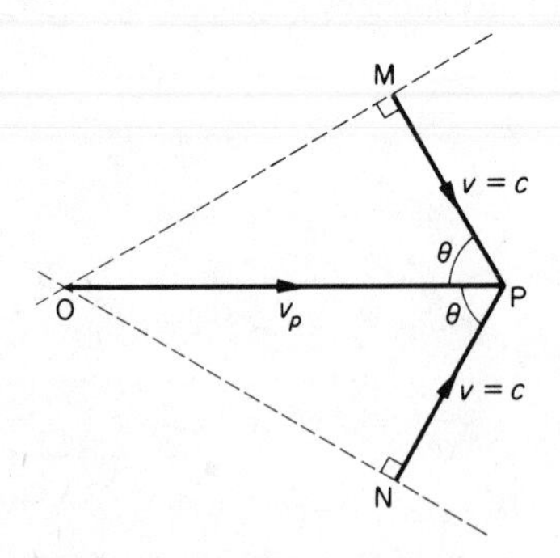

Fig. 27

From the diagram we have

$$v/v_p = \cos\theta$$

or

$$v_p = v/\cos\theta$$

As $\cos\theta < 1$, hence $v_p > c$

If λ is the free space wavelength and λ_g is the guide wavelength we have

$$v = f\lambda$$

$$v_p = f\lambda_g$$

or $$v/v_p = \lambda/\lambda_g = \cos\theta \quad \text{from above}$$

As $$\cos\theta < 1, \quad \text{hence} \quad \lambda_g > \lambda$$

Since $\lambda_g > \lambda$, the composite waveguide pattern travels in the waveguide with a phase velocity $v_p > c$. However, this does not violate the theory of relativity since the energy of the wave is propagated along the waveguide with a group velocity v_g where v_g is the horizontal component of the free space velocity of each component plane wave.

Hence $$v_g = v\cos\theta$$

$$v_p\, v_g = (v/\cos\theta)v\cos\theta = v^2 = c^2$$

or $$v_p\, v_g = c^2$$

5.3 Waveguide equation

Consider the propagation of the H_{10} mode in a rectangular guide with dimensions a, b as shown in Fig. 28.

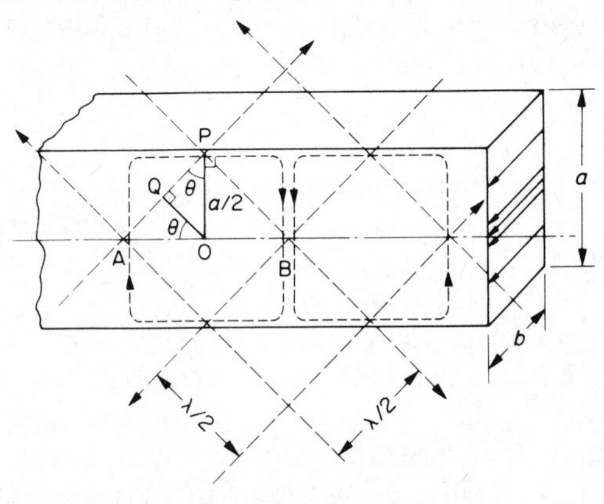

Fig. 28

From the diagram OP and OQ are perpendiculars from the centre point 0 such that

$$OP = a/2$$

$$OQ = \lambda/4$$

From ΔOQP we have

$$\sin\theta = OQ/OP = \frac{\lambda/4}{a/2} = \frac{\lambda}{2a}$$

But $$\cos\theta = v/v_p = (c/f)\lambda_g = \lambda/\lambda_g$$

and $$\sin^2\theta + \cos^2\theta = 1$$

Hence $$(\lambda/2a)^2 + (\lambda/\lambda_g)^2 = 1$$

Now, if λ_c is the cut-off wavelength, $\lambda_c = 2a$ for the $TE_{10}(H_{10})$ mode.*

Hence $$(\lambda/\lambda_c)^2 + (\lambda/\lambda_g)^2 = 1$$

or $$1/\lambda^2 = 1/\lambda_c^2 + 1/\lambda_g^2$$

where λ is the free space wavelength, λ_c is the cut-off wavelength and λ_g is the guide wavelength. This is the basic waveguide equation relating λ, λ_c and λ_g. It can be shown to be true for any mode, either in rectangular or circular waveguide.

EXAMPLE 10

What factors influence the choice of the dimensions for a rectangular waveguide used to transmit an H_{10} wave? Explain the meaning of phase velocity and group velocity in relation to the field-pattern of the wave, and demonstrate from the geometry of the pattern, or otherwise, that the product of these velocities is the square of the velocity of light. Calculate the velocities for this wave at a frequency of 10,000 MHz in a guide which measures 1 in. × $\frac{1}{2}$ in. internally and is air filled.

(L.U. B.Sc.(Eng) Tels. Pt. 3, 1964)

Solution

The dimensions a and b are chosen for propagation of the H_{10} mode at the given frequency f. Since the free space wavelength $\lambda < \lambda_c$ where $\lambda_c = 2a$ for the H_{10} mode, hence $\lambda < 2a$. However, to avoid propagation of the next higher mode H_{20} where $\lambda_c = a$, we must have $\lambda > a$. Hence $\lambda > a > \lambda/2$, the exact value being chosen to allow for $\pm 20\%$ operation on either side of the operating frequency, with minimum attenuation.

Dimension b on the other hand, is not critical, but it determines the power handling capacity of the waveguide since the breakdown voltage across the top and bottom walls of the waveguide, depend upon b. However, b must not be too large or else it will propagate the H_{01} mode. Usually $b = a/2$ in practice, as this gives a value of guide impedance $Z_{TE} = 377\,\lambda_g/\lambda$.

* See Section 5.5.

The phase and group velocities v_p and v_g respectively have been explained in the text in Section 5.2 where it was shown that $v_p v_g = c^2$.

Hence $a = 1 \text{ in.} = 2{\cdot}54 \times 10^{-2}\text{ m}$

$b = \frac{1}{2}\text{ in.} = 1{\cdot}27 \times 10^{-2}\text{ m}$

$\lambda = c/f = (3 \times 10^8)/10^{10} = 3 \times 10^{-2}\text{ m}$

with $\lambda_c = 2a = 5{\cdot}08 \times 10^{-2}\text{ m}$

Since $1/\lambda^2 = 1/\lambda_c^2 + 1/\lambda_g^2$

We have
$$1/\lambda_g^2 = \frac{1}{(3 \times 10^{-2})^2} - \frac{1}{(5{\cdot}08 \times 10^{-2})^2}$$
$$= 0{\cdot}111 \times 10^4 - 0{\cdot}0388 \times 10^4$$
$$= 0{\cdot}723 \times 10^3$$

or $\lambda_g = 3{\cdot}744 \times 10^{-2}\text{ m}$

Now $v_p = f\lambda_g = 10^{10} \times 3{\cdot}744 \times 10^{-2}$

or $v_p = 3{\cdot}744 \times 10^8\text{ m/s}$

with
$$v_g = c^2/v_p = \frac{9 \times 10^{16}}{3{\cdot}744 \times 10^8}$$

or $v_g = 2{\cdot}4 \times 10^8\text{ m/s}$

5.4 Rectangular waveguides

The properties of waveguides may be studied by solving Maxwell's equations for propagation, in bounded regions. The solutions to Maxwell's equations for rectangular and circular waveguides are obtained by satisfying the necessary boundary conditions for propagation within the waveguide.

The general equation for wave propagation in space is

$$\nabla^2\psi = \frac{1}{c^2}\frac{\partial^2\psi}{\partial t^2}$$

where ψ is the function **E** or **H**.

Since we are essentially concerned with alternating quantities, which may be represented by $e^{j\omega t}$, hence $\partial^2/\partial t^2 \equiv -\omega^2$.

with
$$\nabla^2\psi = -(\omega^2/c^2)\psi$$

or
$$\nabla^2\psi = -k^2\psi$$

where $k = \omega/c$ gives the eigen-values of the system. Solutions to this equation are given in Jordan.[15]

5.5 Rectangular modes

Two types of guided waves are now possible, namely, transverse electric (TE) wave for which $E_z = 0$ and transverse magnetic (TM) wave, for which $H_z = 0$. By using Maxwell's equations for curl **E** and curl **H**, with the solutions for E_z and H_z, the general field components for the various modes can be obtained. Further details are given in Jordan,[15] and the relevant field components are given hereafter.

TE modes

$$E_x = \frac{j\omega\mu H_0}{k_c^2}\frac{n\pi}{b}\cos\frac{m\pi x}{a}\sin\frac{n\pi y}{b}\,e^{j(\omega t-\beta z)}$$

$$E_y = \frac{-j\omega\mu H_0}{k_c^2}\frac{m\pi}{a}\sin\frac{m\pi x}{a}\cos\frac{n\pi y}{b}\,e^{j(\omega t-\beta z)}$$

$$E_z = 0$$

$$H_x = \frac{\gamma H_0}{k_c^2}\frac{m\pi}{a}\sin\frac{m\pi x}{a}\cos\frac{n\pi y}{b}\,e^{j(\omega t-\beta z)}$$

$$H_y = \frac{\gamma H_0}{k_c^2}\frac{n\pi}{b}\cos\frac{m\pi x}{a}\sin\frac{n\pi y}{b}\,e^{j(\omega t-\beta z)}$$

$$H_z = H_0\cos\frac{m\pi x}{a}\cos\frac{n\pi y}{b}\,e^{j(\omega t-\beta z)}$$

The simplest TE mode is $TE_{10}(H_{10})$ or dominant mode which has the longest cut-off wavelength. For the TE_{10} mode this is given by

$$\lambda_c = 2/\sqrt{(1/a)^2} = 2a$$

The field components for the TE_{10} mode are:

$$E_x = 0$$

$$E_y = A\sin\frac{\pi x}{a}\,e^{j(\omega t-\beta z)}$$

$$E_z = 0$$

with $$A = \frac{-j\omega\mu H_0}{k_c^2}\pi/a$$

$$H_x = B\sin\frac{\pi x}{a}e^{j(\omega t-\beta z)}$$

$$H_y = 0$$

$$H_z = H_0\cos\frac{\pi x}{a}e^{j(\omega t-\beta z)}$$

with $$B = \frac{\gamma H_0}{k_c^2}\pi/a$$

Hence E_y has a half sine wave distribution over the waveguide cross-section with a peak value at the centre of the cross-section where $x = a/2$.

Also $$E_y/-H_x = A/-B = \frac{\omega\mu}{\beta} = (\omega/c)\mu c(1/\beta) = Z_0(\lambda_g/\lambda)$$

where $Z_0 = \sqrt{\mu_0/\varepsilon_0}$ is the intrinsic impedance of free space. The ratio $E_y/-H_x$ is *defined* as the impedance Z_{TE} for the TE wave.

Hence $$Z_{TE} = Z_0(\lambda_g/\lambda) = 377\lambda_g/\lambda \text{ ohms}$$

TM modes

$$E_x = \frac{-\gamma E_0}{k_c^2}\frac{m\pi}{a}\cos\frac{m\pi x}{a}\sin\frac{n\pi y}{b}e^{j(\omega t-\beta z)}$$

$$E_y = \frac{-\gamma E_0}{k_c^2}\frac{n\pi}{b}\sin\frac{m\pi x}{a}\cos\frac{n\pi y}{b}e^{j(\omega t-\beta z)}$$

$$E_z = E_0\sin\frac{m\pi x}{a}\sin\frac{n\pi y}{b}e^{j(\omega t-\beta z)}$$

$$H_x = \frac{j\omega\varepsilon E_0}{k_c^2}\frac{n\pi}{b}\sin\frac{m\pi x}{a}\cos\frac{n\pi y}{b}e^{j(\omega t-\beta z)}$$

$$H_y = \frac{-j\omega\varepsilon E_0}{k_c^2}\frac{m\pi}{a}\cos\frac{m\pi x}{a}\sin\frac{n\pi y}{b}e^{j(\omega t-\beta z)}$$

$$H_z = 0$$

Similarly, for the TM modes we have

$$E_y/-H_x = \beta/\omega\varepsilon = \frac{\beta c}{\omega\varepsilon c} = \frac{2\pi}{\lambda g} \times \frac{\lambda}{2\pi} \times \frac{1}{\varepsilon c} = Z_0(\lambda/\lambda_g)$$

which is *defined* as the impedance Z_{TM}.

Hence $$Z_{TM} = Z_0(\lambda/\lambda_g) = 377\lambda/\lambda_g \text{ ohms}$$

The simplest TM mode is the TM_{11} or dominant mode since TM modes vanish when $n = 0$. Typical field patterns for some rectangular modes are shown in Fig. 29.

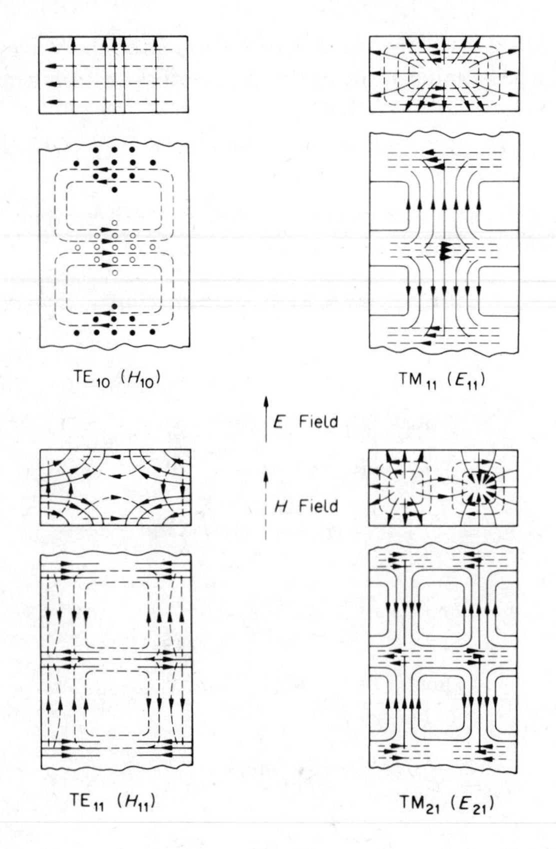

Fig. 29

EXAMPLE 11

A rectangular waveguide has internal dimensions a and b as shown in Fig. 30. A mode has the electric field given by

$$E_x = A \sin \frac{\pi y}{b} \exp\left[-j\beta z + \omega t\right]$$

$$E_y = E_z = 0$$

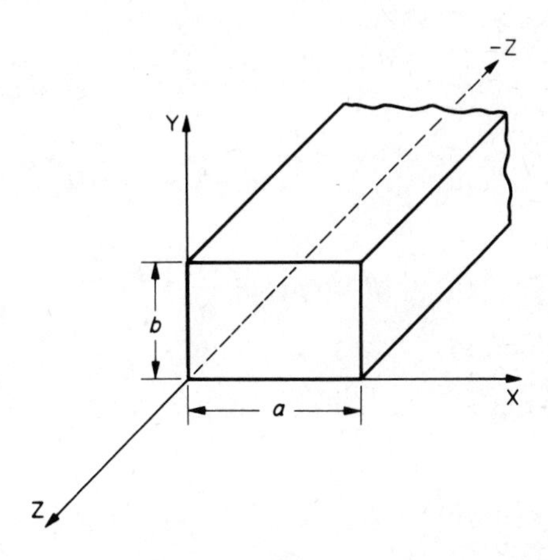

Fig. 30

Derive an expression for the phase-change coefficient β in terms of a, b and the electrical constants of the mediums filling the guide. Obtain the corresponding magnetic field.

Describe how the guide may be coupled to a coaxial line so that the above mode is excited within the guide.

(C.E.I. Pt. 2, Comm. Eng., 1968)

Solution

The mode propagated is the H_{01} where $m = 0, n = 1$

Now
$$\gamma = j\beta$$

$$\partial/\partial z \equiv -\gamma$$

$$\partial/\partial t \equiv j\omega$$

Also
$$\text{curl } \mathbf{E} = -\frac{\partial \mathbf{B}}{\partial t} = -j\omega\mu\mathbf{H}$$

Hence
$$\frac{\partial E_z}{\partial y} - \frac{\partial E_y}{\partial z} = -j\omega\mu H_x$$

$$-\frac{\partial E_z}{\partial x} + \frac{\partial E_x}{\partial z} = -j\omega\mu H_y$$

$$\frac{\partial E_y}{\partial x} - \frac{\partial E_x}{\partial y} = -j\omega\mu H_z$$

Since $E_Z = 0$, we obtain
$$\frac{\partial E_x}{\partial z} = -j\omega\mu H_y$$

or
$$-j\beta A \sin\frac{\pi y}{b}\, e^{-\gamma z} e^{j\omega t} = -j\omega\mu H_y$$

or
$$H_y = (\beta A/\omega\mu)\sin\frac{\pi y}{b}\, e^{-\gamma z} e^{j\omega t}$$

Since $E_y = 0$, we obtain
$$-\frac{\partial E_x}{\partial y} = -j\omega\mu H_z$$

or
$$-(\pi/b)A\cos\frac{\pi y}{b}\, e^{-\gamma z} e^{j\omega t} = -j\omega\mu H_z$$

or
$$H_z = -(j\pi A/\omega\mu b)\cos\frac{\pi y}{b}\, e^{-\gamma z} e^{j\omega t}$$

Also
$$1/\lambda_g^2 = 1/\lambda^2 - 1/\lambda_c^2$$

with
$$\lambda_c = \frac{2}{\sqrt{0 + (1/b)^2}} = 2b$$

$$\omega/c = 2\pi/\lambda$$

$$\beta = 2\pi/\lambda g$$

Hence
$$(\beta/2\pi)^2 = (\omega/2\pi c)^2 - (1/2b)^2$$

or
$$\beta = \sqrt{\omega^2\mu\varepsilon - \pi^2/b^2}$$

The H_{10} mode may be launched by inserting a $\lambda/4$ length of coaxial cable into the side of the guide, placed $\lambda g/4$ from a moveable piston as shown in Fig. 31. For the H_{01} mode the cable enters side b.

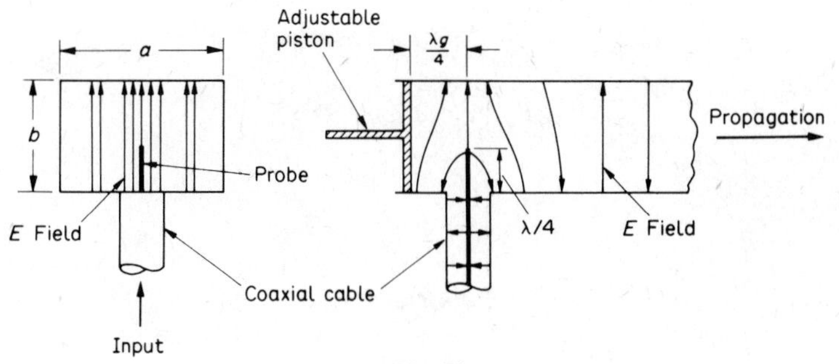

Fig. 31

5.6 Circular waveguides

The solution of the wave equation for circular waveguides is conveniently obtained using cylindrical coordinates. Hence we have

$$\nabla^2 \psi = -(\omega^2/c^2)\psi$$

where $\psi = \mathbf{E} \text{ or } \mathbf{H}$

and $$\nabla^2 = \partial^2/\partial r^2 + \frac{1}{r}\partial/\partial r + \frac{1}{r^2}\partial^2/\partial\phi^2 + \partial^2/\partial z^2$$

in cylindrical coordinates. Further details are given in Ramo.[16]

5.7 Circular modes

The general field equations for these modes are given below and are obtained through Maxwell's curl equations, together with the solutions for **E** and **H**. Further details are given by Ramo.[16]

TE$_{mn}$ modes

$$E_r = \frac{-\omega\mu m H_0}{r k_c^2} J_m(k_c r)\cos m\phi \mathrm{e}^{\mathrm{j}(\omega t - \beta z)}$$

$$E_\phi = \frac{\mathrm{j}\omega\mu}{k_c} H_0 J_m'(k_c r)\cos m\phi \mathrm{e}^{\mathrm{j}(\omega t - \beta z)}$$

$$E_z = 0$$

$$H_r = \frac{-\gamma H_0}{k_c} J_m'(k_c r)\cos m\phi \mathrm{e}^{\mathrm{j}(\omega t - \beta z)}$$

$$H_\phi = \frac{-\beta m H_0}{r k_c^2} J_m(k_c r)\cos m\phi e^{j(\omega t - \beta z)}$$

$$H_z = H_0 J_m(k_c r)\cos m\phi e^{j(\omega t - \beta z)}$$

TM_{mn} modes

$$E_r = \frac{-\gamma E_0}{k_c} J'_m(k_c r)\cos m\phi e^{j(\omega t - \beta z)}$$

$$E_\phi = \frac{-\beta m E_0}{r k_c^2} J_m(k_c r)\cos m\phi e^{j(\omega t - \beta z)}$$

$$E_z = E_0 J_m(k_c r)\cos m\phi e^{j(\omega t - \beta z)}$$

$$H_r = \frac{\omega \varepsilon m E_0}{r k_c^2} J_m(k_c r)\cos m\phi e^{j(\omega t - \beta z)}$$

$$H_\phi = \frac{-j\omega\varepsilon E_0}{k_c} J'_m(k_c r)\cos m\phi e^{j(\omega t - \beta z)}$$

$$H_z = 0$$

Comments

1. $E_r/H_\phi = \omega\mu/\beta = Z_{TE}$ for TE modes
2. $E_r/H_\phi = \gamma/j\omega\varepsilon = Z_{TM}$ for TM modes

Typical field patterns for some TE and TM modes are shown in Fig. 32.

5.8 Waveguide attenuation

As electromagnetic waves travel down the waveguide, currents are induced in the walls of the waveguide which give rise to losses, and so the wave is gradually attenuated as it travels along the waveguide. The attenuation is a function of the waveguide material and dimensions, the wavelength and mode of propagation. To reduce losses the inner surface of the waveguides may be coated with silver or gold plating.

Calculations of losses can be made from theoretical reasoning using the field equations. Typical curves are given in Fig. 33 from which it will be seen that losses are greater for higher order modes. They are also higher for circular as compared to rectangular modes. However, the

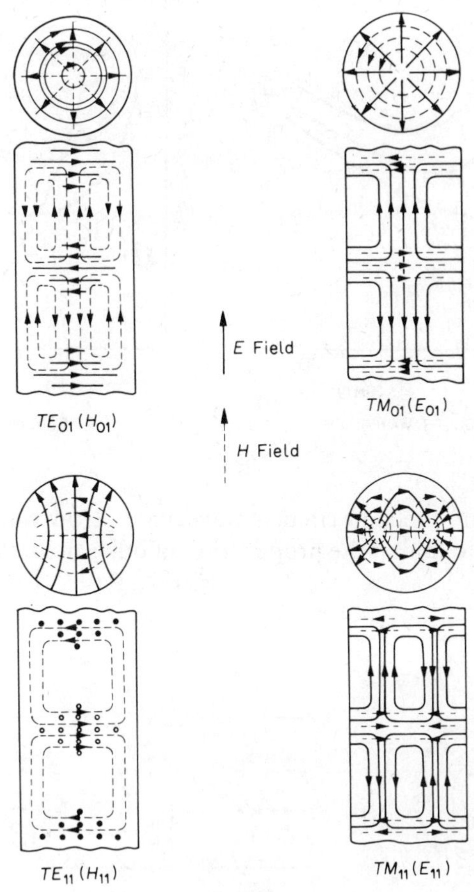

Fig. 32

$TE_{01}(H_{01})$ circular mode appears to have very low losses and in fact it tends to zero at higher frequencies. In practice, circular-mode losses are greater than the theoretical values because of the difficulty of making a perfectly circular guide. Recent work[8,9] has enabled the TE_{01} to be exploited for long-distance trunk communication by using specially constructed low-loss circular waveguide.

EXAMPLE 12

Draw the field patterns for the H_{01} mode in a circular waveguide and explain why this mode is chosen for trunk telecommunications applications. Estimate the critical wavelength for this mode and show that it is

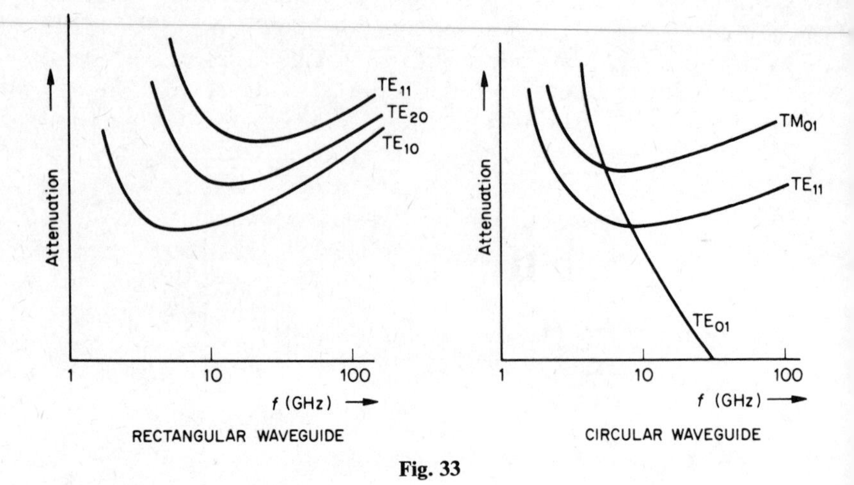

Fig. 33

not the dominant mode in circular waveguide. How can the waveguide be constructed to inhibit the propagation of other modes?

(C.E.I., Part 2, Comm. Eng. 1971)

Solution

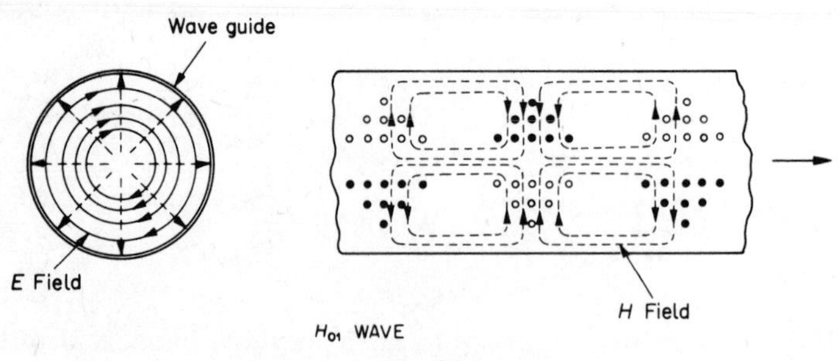

Fig. 34

The H_{01} mode is chosen for long-distance trunk communications because it has very low losses which tend to zero at high frequencies. Typically, the attenuation at 16 GHz is about 0·005 dB/m in a 5 cm diameter copper wall circular waveguide, which is the lowest attenuation for any circular mode.

The critical wavelength for this mode is $1{\cdot}64a$, where a is the radius of the waveguide. By using a larger radius (oversize waveguide) than that given by the cut-off condition, attenuation can be made small and it becomes attractive for long-distance communication. On the other hand, the dominant circular mode is the $TE_{11}(H_{11})$ with the longest cut-off wavelength of $3{\cdot}42a$, but its losses are much greater.

However, a disadvantage of oversize waveguide is the propagation of spurious modes with consequent loss of power. Recent work undertaken has led to the suppression of unwanted modes by making them travel more slowly along the waveguide than the H_{01} mode, thus reducing energy transfer to these modes. This is done by using either a circular waveguide made of a wire helix or a metal surface which has corrugations or a thin coating of dielectric material. Mode filters may also be employed to clean up any remaining unwanted modes.

5.9 Launching in waveguides

This can usually be done from a coaxial cable. Electrostatic coupling (Fig. 35) employs a short probe which behaves as an antenna, while electromagnetic coupling is possible by using a small closed loop.

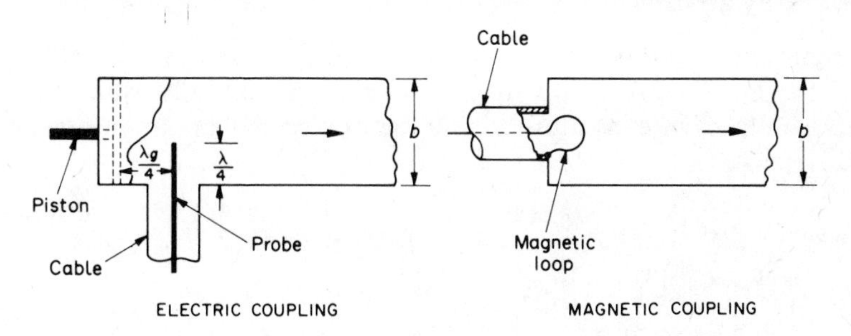

Fig. 35

By inserting the probe $\lambda/4$ into the waveguide and $\lambda g/4$ from one end of the waveguide (the position being adjustable by a piston), waves are propagated towards the right. The field pattern near the probe contains higher order modes but further down the waveguide, a single mode is transmitted.

In the case of magnetic coupling (Fig. 35), the size of the loop is not critical and it couples in the magnetic field lines. The end of the loop is terminated on the waveguide wall and behaves as a single turn coil antenna.

6
Microwave techniques

The increasing use of microwaves because of demands for higher frequencies due to the congestion at lower frequency bands, has led to considerable development in this field. Microwave engineering is now a firmly established branch of Telecommunications and a broad knowledge of the various aspects pertinent to this field are essential. These aspects may be broadly classified under sources, components and measurements.

6.1 Microwave sources[17]

The main source of microwave energy in the past has been the klystron oscillator. Two basic forms are the two-cavity klystron and the reflex klystron, the latter being the most common in use.

Two-cavity klystron

It comprises two tuned cavities called the buncher and catcher, as shown in Fig. 36. Electrons are made to bunch periodically while passing between the cavities as illustrated in the Applegate diagram in Fig. 36.

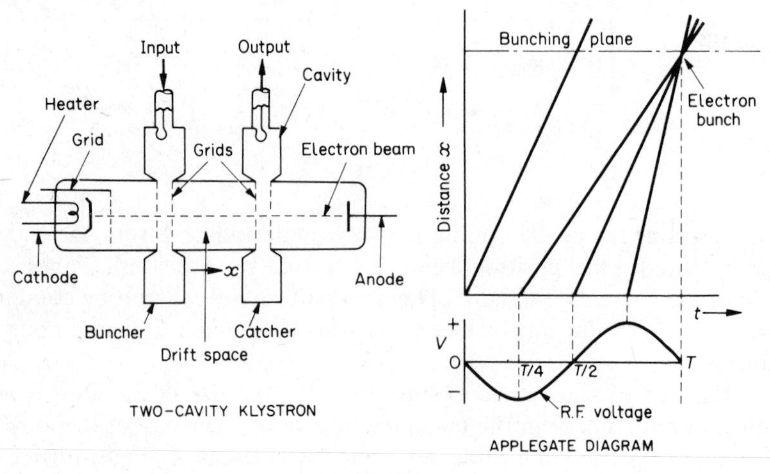

Fig. 36

The stream of electrons emitted from a cathode are attracted to the anode by an accelerating d.c. potential. Assuming an r.f. voltage exists across the first cavity, as the electrons pass through it, some are accelerated during the positive half cycle and others are decelerated during the negative half cycle. Hence the uniform stream of electrons is broken up into bunches of electrons as they drift forward towards the second cavity. This bunching gives rise to current pulses which have a frequency component equal to the r.f. voltage across the first cavity.

On approaching the catcher, the bunches give up energy to it if they arrive in such a phase as to be decelerated. The r.f. voltage across the catcher builds up while the bunches of electrons are collected by the anode. Such a device is basically an amplifier, but if part of the output of the catcher is fed back to the buncher in the right phase, an oscillator is obtained.

Reflex klystron

It is a microwave oscillator which uses only one resonant cavity as shown in Fig. 37.

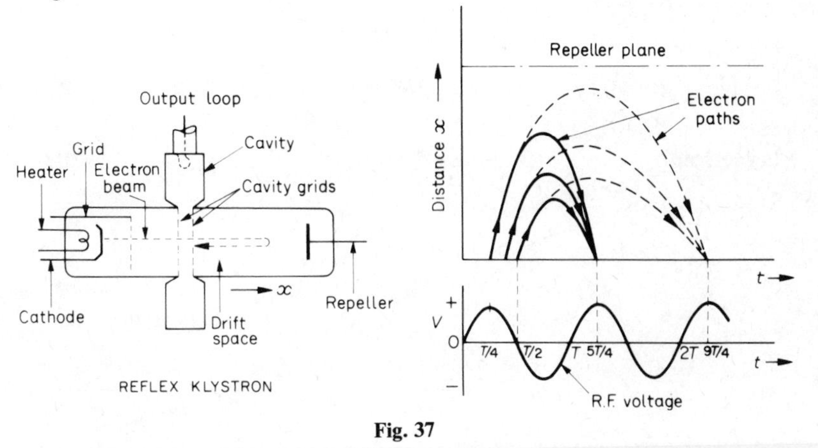

Fig. 37

The stream of electrons after being accelerated through the resonant cavity are then repelled back by the repeller which is at a negative potential to the cathode. The electrons on returning form bunches which move back through the cavity and give up energy to the cavity if they return at the right phase of the accelerating R.F. cycles, as shown in Fig. 37.

The oscillating condition can be established at various values of repeller voltage and are known as its modes. The lowest mode corresponding to the shortest bunching time and highest (negative) repeller voltage is the $\frac{3}{4}$ mode, while the other modes are known as the $1\frac{3}{4}$, $2\frac{3}{4}$, etc. Most power may be coupled out of the $1\frac{3}{4}$ mode as normally power cannot be coupled out of the $\frac{3}{4}$ mode due to the conductance of the system.[18] This is shown in Fig. 38.

The reflex klystron can be tuned mechanically over a small range of frequency by an external screw mechanism which alters the cavity size.

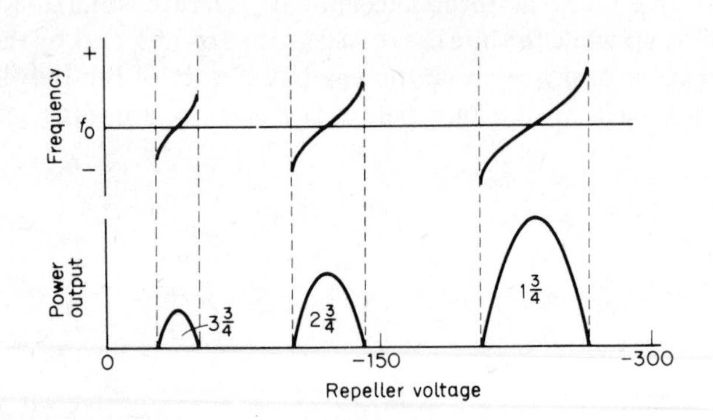

Fig. 38

Magnetron

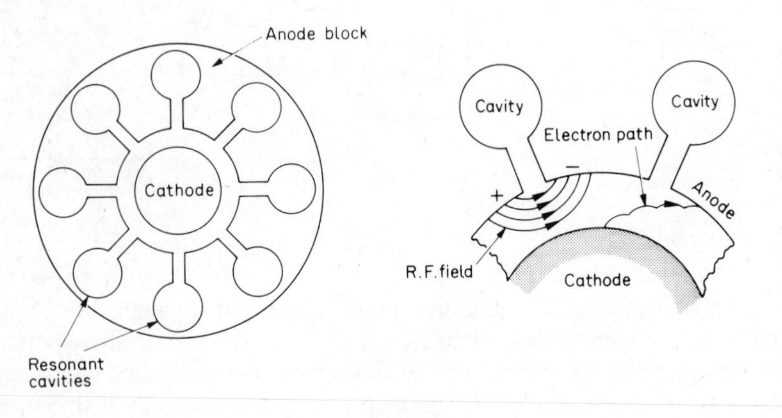

Fig. 39

A source of high-power microwave energy is the multi-cavity magnetron shown in Fig. 39. It is basically a cylindrical diode with the cathode at the centre and an anode composed of several resonant cavities placed around it. A steady magnetic field is applied parallel to the cathode and a d.c. or pulsed voltage is applied between anode and cathode. The electrons from the cathode follow cycloidal paths to the anode and their d.c. energy is converted to r.f. energy at the cavities.

The most normal mode of operation is the π-mode, when alternate segments are in phase. If the electron motion is correct, it will reach the anode segments with little d.c. energy as it will have lost all its energy to the r.f. field across the oscillating cavities. To ensure the π-mode operation, alternate segments are joined by metal straps called *strapping*.

The exact build up of oscillations is not clear but is believed to be associated with the instability of the space charge cloud around the cathode, which rotates and breaks up into spoke-like arms reaching out to the anode block. However, high efficiencies around 60% are possible and due to back bombardment of the cathode, the heaters may be switched off during operation, to avoid overheating.

The performance of a magnetron may be studied by means of a performance chart and a Rieke diagram shown in Fig. 40. The performance chart shows how the applied voltage and current vary, for various values of applied field at a particular load impedance, while the Rieke diagram shows how the power output varies with frequency of operation under

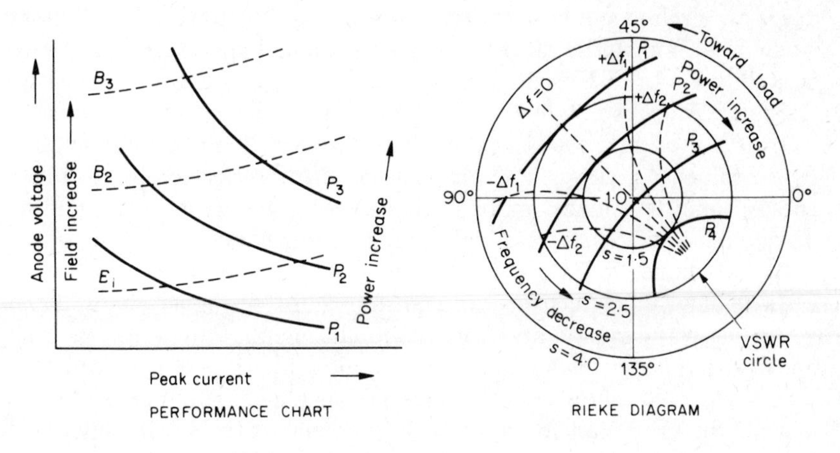

Fig. 40

various load conditions. The latter is essentially a Smith Chart with the VSWR circles drawn in, while the reactance circles are omitted. In this connection, the *pushing* figure of a magnetron is the change of frequency for a defined anode voltage variation at a given load, which is obtained from the performance chart, and the *pulling* figure is the change of frequency for a given change in the standing wave ratio at the output of the magnetron and is obtained from the Rieke diagram.

Gunn diode oscillator

In recent years, a more convenient form of microwave oscillator which uses a low-voltage d.c. supply instead of the cumbersome high voltage supply of the klystron, is the Gunn diode oscillator.

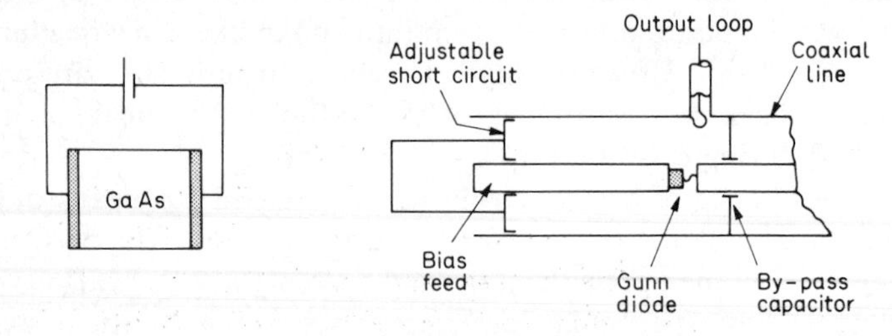

Fig. 41

This is a semiconductor device shown in Fig. 41, in which oscillations occur in bulk material rather than across a junction. The material used is mainly GaAs and oscillations are due to the existence of a negative resistance which is obtained on applying a d.c. voltage to the material.

The Gunn effect requires the presence of two conduction bands separated by a small energy gap as in GaAs. The conduction electrons in the lower energy band have higher mobility than those in the upper band, but the density of allowable states is higher in the upper band.

At low applied fields, the current increases linearly with voltage but at a threshold value, electrons are transferred to the higher energy band which has a lower mobility. Lower mobility implies lower current with increasing voltage which amounts to a region of negative resistance.

It is observed that part of the material has a high field gradient across it while the other part has a low field gradient across it. The high field region or *domain* travels from the cathode with a drift velocity to the end of the material and collapses at the anode, while another *domain* is generated at the cathode. The sudden collapse of the *domain* generates a current pulse and if the transit time through the material is such that the current pulses occur at a microwave frequency, microwave power generation is possible in an external circuit. For this to occur, the GaAs sample must be very thin (about 10^{-3} cm) and it must be placed inside a resonant cavity which may be tuned.

D.C. power is thus converted directly into microwave energy and the efficiency is about 1%. The low applied voltage of about 6 volts and the small overall size of the device make it very attractive as a microwave oscillator.

6.2 Microwave components[19,20]

The quality and precision of microwave components is of prime importance in making accurate measurements and it depends on good engineering practice and the tolerances employed. Enumerated below are some typical waveguide components with a short description of each.

Bends and Twists

Most waveguides are manufactured in straight sections. However, to go around corners or bends, *E*-plane or *H*-plane bends are used over short lengths. Furthermore, by means of a twist, the plane of polarisation can be changed from horizontal to vertical.

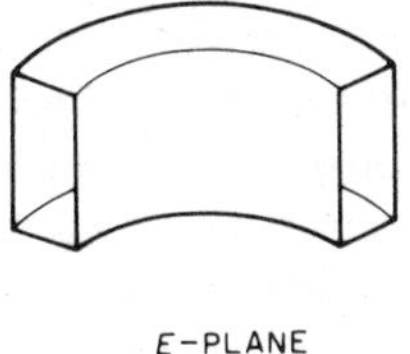

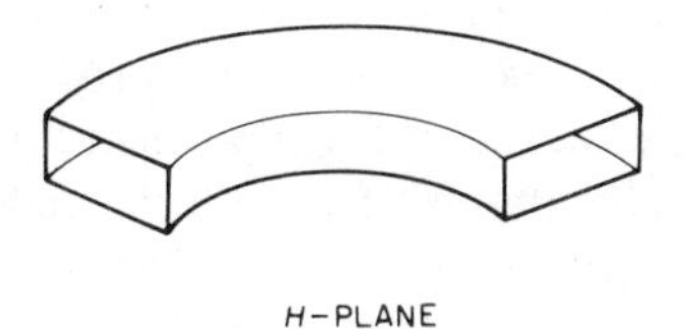

Fig. 42

T-junctions

To provide additional branches, *T*-junctions are used. These are either of the *E*-plane or *H*-plane construction or a combination of both which is known as the hybrid or magic *T*. In the case of a *T*-junction with matched arms, power entering arm (1) divides equally into arms (2) and (3) and is a three-port network. The hybrid *T* is a four-port network and is such that power into port C divides equally into arms A and B, while power into arm D divides equally but in antiphase in arms A and B. Ports C and D are thus completely isolated as shown in Fig. 48.

A variation of the four-port *T*-junction is the hybrid ring which consists of four *T*-junctions arranged in a ring. Power into (1) appears at two others and the third one is isolated as shown in Fig. 43.

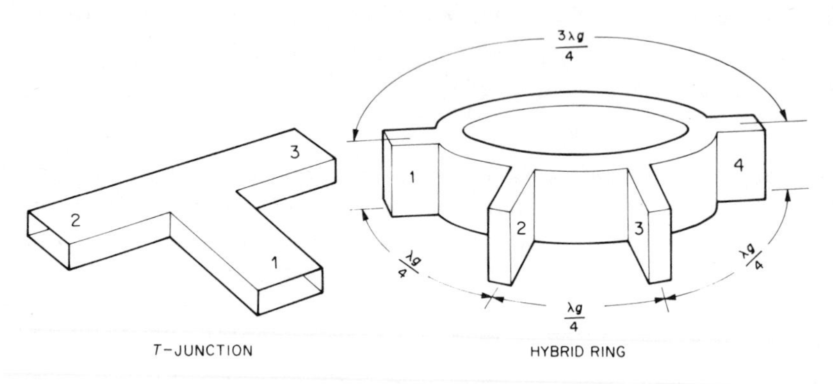

Fig. 43

Directional coupler

This is a matched four-port device, power being fed into one and obtained out of two others. Two important properties are the coupling and directivity. For a typical four arm coupler with matched arms shown in Fig. 44, power in at (1) appears at (2) and (4) and power in at (2) appears at (3). The amount from (2) depends on the coupling which may vary from 3 dB to 60 dB. Coupling may be through slots in the precision types. The basic idea of operation depends on the distances between the holes which are such that waves in the forward direction are in phase, while those in the opposite direction cancel out. For a four-arm coupler with power P_1 into port (1) and power P_3 and P_4 from ports (3) and (4) respectively, the coupling and directivity are defined by

$$\text{Coupling} = 10 \log P_1/P_4 \text{ dB}$$

$$\text{Directivity} = 10 \log P_4/P_3 \text{ dB}$$

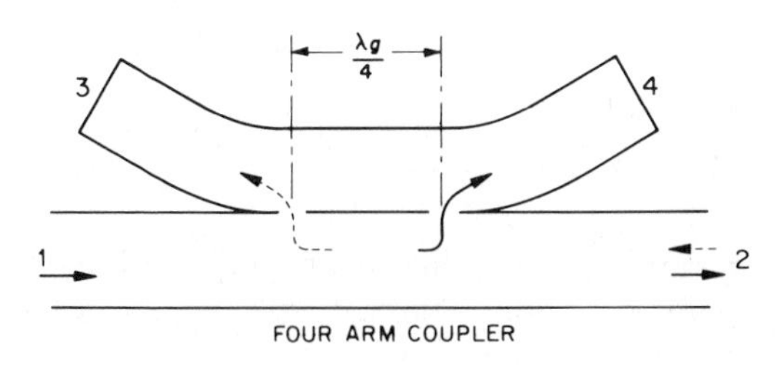

Fig. 44

Attenuators

For general purpose use, vane attenuators are used to attenuate power flowing down the guide. They consist of a glass vane coated with a resistive material like Nickel or Chromium which is inserted in the path of the wave to a variable depth in the waveguide, by a screw mechanism. The greater the depth, the greater the attenuation and it may be calibrated in dB.

For higher precision a rotary vane attentuator is used. It comprises a central circular piece of waveguide with rectangular ends, each with a resistance vane and the central part is rotatable as shown in Fig. 45.

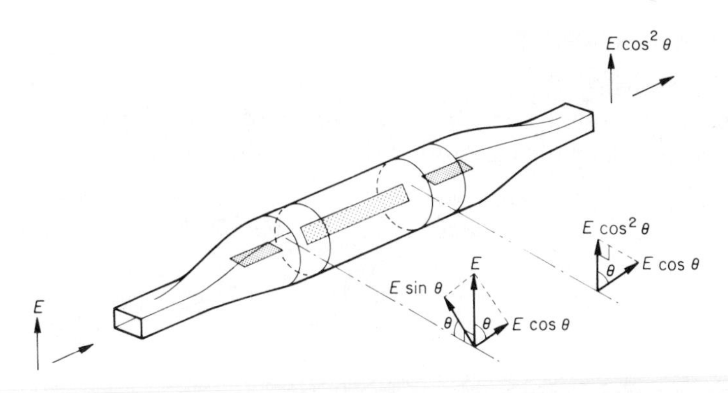

Fig. 45

Resolving the E vector along and perpendicular to the vane, the component $E\sin\theta$ is attenuated while $E\cos\theta$ is transmitted and at the next plate this becomes $E\cos^2\theta$ and is further transmitted.

Hence, the power finally transmitted is proportional to $\cos^2 \theta$ and the attenuation in dB is given by

$$\mathrm{dB} = 20 \log_{10} \left[\frac{E}{E \cos^2 \theta} \right] = 40 \log_{10} \sec \theta$$

The angular rotation can be calibrated directly in dB with little backlash and with an accuracy of about 0·005 dB.

Wavemeters

These are of two kinds—the transmission wavemeter and the reaction wavemeter. The reaction type is generally used since it is placed across the waveguide and can be isolated without dismantling it, by simply detuning it. When tuned to the signal frequency, it absorbs power and the power absorption is easily observed on a power meter.

Operation is usually in the TE_{011} mode* comprising (a) a waveguide resonant cavity (b) a tuning plunger (c) a driving screw mechanism. The resonant cavity is precision machined, cylindrical and gold-flashed for a high Q. It is mounted on a piece of rectangular waveguide and the cavity size is altered by moving the tuning plunger which is attached to the screw mechanism.

Standing wave detector

This is probably the most important and expensive component in a waveguide set-up. Essentially, it is a section of waveguide with a narrow horizontal slot on top as shown in Fig. 46. Mounted above is a moving

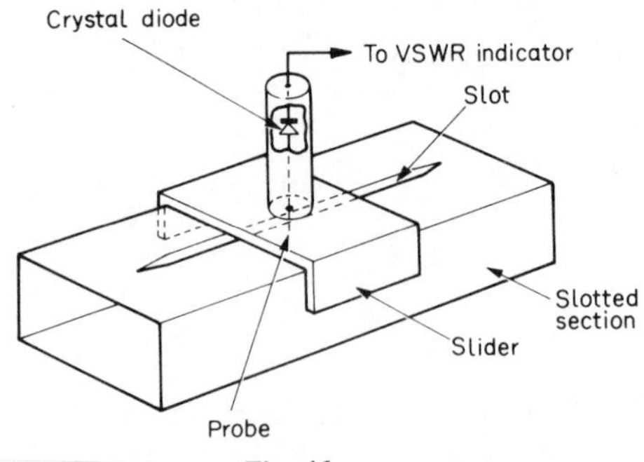

Fig. 46

* See Appendix C.

mechanism with a vertical adjustable probe inserted into the slot. The whole unit is extremely accurately machined and carefully mounted for precision.

It is used to measure the E field along the slot over a few wavelengths and from this the VSWR, reflection coefficient and mismatch of the line can be ascertained.

Ferrite components

Ferrites are ferrimagnetic substances which possess non-reciprocal properties. This is due to spinning electrons which are loosely bound to the parent atoms. On applying an external d.c. field, the spinning electrons behave as little dipole magnets and precess about the applied magnetic field with a Larmor frequency ω given by $\omega = \gamma H$, where γ is the gyromagnetic ratio and H is the applied field. This is known as the gyromagnetic effect and energy can be coupled to the spinning electrons from an external microwave field to provide the energy of precession, thereby causing a resonance loss of microwave energy.

Two important microwave components using ferrites are the *isolator* and *circulator* shown in Fig. 47.

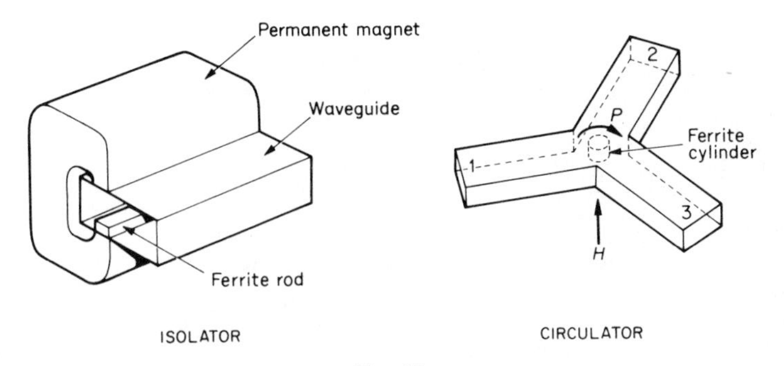

Fig. 47

An isolator is a device which has a low loss in the forward direction but considerable loss in the reverse direction. In one form of resonance isolator, a piece of ferrite is placed asymmetrically inside a rectangular waveguide. On applying a suitable magnetic field by means of a permanent magnet, a loss of 0·5 dB occurs in one direction as against 30 dB in the reverse direction. It is used for isolating an oscillator from its load, as far as reflections are concerned, thereby producing a 'matched' condition.

A circulator is similarly a device for allowing power to circulate in one direction only, e.g. clockwise. A typical circulator comprises three or more waveguide arms meeting at a junction. A ferrite post is placed at the centre and a vertical magnetic field is applied in one direction. For a three-port circulator, power may be fed into port 1 which flows into an antenna coupled to port 2. Any reflected power travels to port 3 only, where it is absorbed in a matched load.

EXAMPLE 13

A hybrid T-junction consists of a length of rectangular waveguide AB with a shunt limb C and a series limb D. Draw a sketch of this device and label each limb.

Assuming that A and B are terminated in reflectionless loads, explain by means of separate diagrams showing the *E*-field patterns, why no energy emerges

(a) from D, when the hybrid is energised via limb C;

(b) from C, when the hybrid is energised via limb D.

(C. & G. Adv. Mic. Principles 1969)

Solution

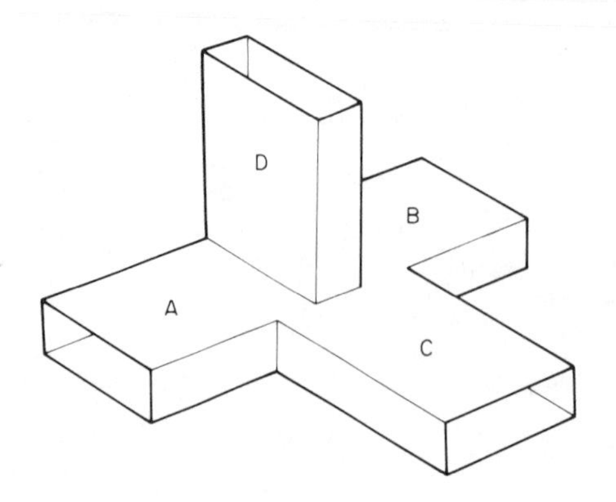

Fig. 48

(a) Power fed into limb C divides equally into limbs A and B due to symmetry, and are in phase. No power therefore emerges from limb D which is thus isolated from limb C. This is shown in Fig. 49(a).

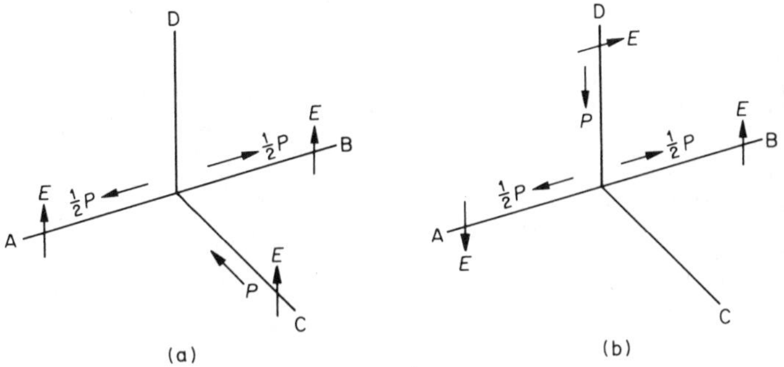

Fig. 49

(b) Power fed into limb D divides equally into limbs A and B due to symmetry, but in opposite phase. No power therefore emerges from limb C which is thus isolated from limb D. This is shown in Fig. 49(b).

6.3 Microwave measurements[21,22]

In the manufacture, installation and operation of microwave equipment it is necessary to make precise measurements at microwave frequencies. Such measurements are somewhat more difficult than those at radio frequencies, because microwave frequencies are in the gigahertz range which for X-band, lies between 8·2 GHz to 12·4 GHz.

Such measurements usually involve a typical test-bench set-up—a grade 2 for general purpose measurements and a grade 1 for precision measurements. In the past, it was usual to make measurements at spot frequencies which generally used narrow band equipment. More recently, however, swept frequency techniques are being used, which call for broad-band characteristics. The measurements may also be made with the aid of a computer which is able to store data and program the measurements. A typical waveguide test-bench set-up is shown in Fig. 50.

The source of microwave power is a klystron or solid state source usually at X-band. This is followed by an isolator which prevents any reflections from affecting the klystron stability. The attenuator (general purpose or rotary type) sets the power level in the guide to a suitable value and the wavemeter is tuned to check the frequency and then detuned. A directional coupler may be used to divide the power into two branches and a known proportion of the incident power such as half of it (3 dB

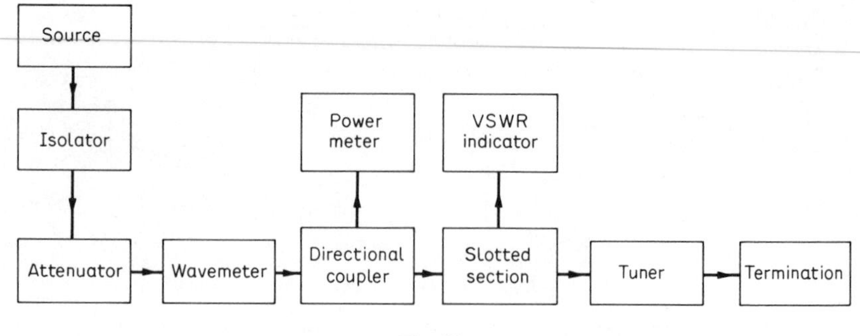

Fig. 50

coupler), is coupled to one arm and monitored by a suitable power meter.

The rest of the power passes to the slotted line which keeps a check on the VSWR. The output of the slotted line probe is fed into the VSWR indicator which reads VSWR directly. The power down the main waveguide is then fed via a three stub tuner and ends in the termination, which may be a short-circuit, an open-circuit or matched termination. The three stub tuner enables better matching to be achieved when required.

Frequency measurements

The measurement of a microwave frequency often requires great accuracy. Frequency measuring devices are either mechanical devices in which physical dimensions are used to determine frequencies such as cavity wavemeters and slotted lines or electronic devices, in which frequency is determined electronically, using heterodyne wavemeters and frequency counters.

An indirect way of measuring frequency is by first measuring wavelength with the use of a slotted line. The standing wave pattern consists of maxima and minima due to the combination of incident and reflected waves. The distance between two adjacent maxima or minima is $\lambda g/2$. This can be converted to frequency using a standard conversion factor.

A more accurate means of measuring frequency is with the use of a resonant cavity or cavity wavemeter. The cavity volume is varied and can be calibrated directly against frequency or given on a calibration chart. Cavity wavemeters have a high Q and give excellent accuracies between 0·005% to 0·5% if temperature compensated by special construction.

High accuracy frequency measurements can be made with the use of a frequency counter to measure the result of two heterodyned signals. The standard signal used is generated harmonically from a crystal controlled oscillator and is mixed with the microwave signal. The accuracy of the measurement is determined by the crystal stability and the accuracy of the frequency counter.

Impedance measurements

At microwave frequencies, impedance is difficult to measure directly and is not used very often. Moreover, its value in general depends at which point in the waveguide the measurement is made. Since impedance impedes the flow of energy, it can be measured by considering the transmission characteristics of the propagated energy. Reflections due to mismatches, set up a standing wave pattern in the waveguide which can be measured or alternatively, the incident and reflected powers can be compared. Consequently, the standing wave ratio or reflection coefficient are the parameters usually measured and from either of them, the impedance can be obtained.

The measurement of the voltage standing wave ratio (VSWR) is accomplished with the use of a slotted section of waveguide wherein the pattern of electrical energy can be probed. When the effective impedance of a waveguide matches that of the source, no reflections occur and the VSWR is effectively unity. If an impedance mismatch is present the incident and reflected waves combine vectorially to give a VSWR different from unity.

Measuring the intensity of the incident and reflected waves provides another means of obtaining the same result. This method requires an isolating device called a directional coupler with a high directivity for sampling the incident and reflected waves. The ratio of reflected to incident power is the reflection coefficient and can be measured by the reflectometer method.

VSWR measurement

Low VSWR

The incident and reflected waves combine to produce maxima and minima as shown in Fig. 51. The VSWR has been defined as

$$\text{VSWR} = \frac{V_{\text{max}}}{V_{\text{min}}}$$

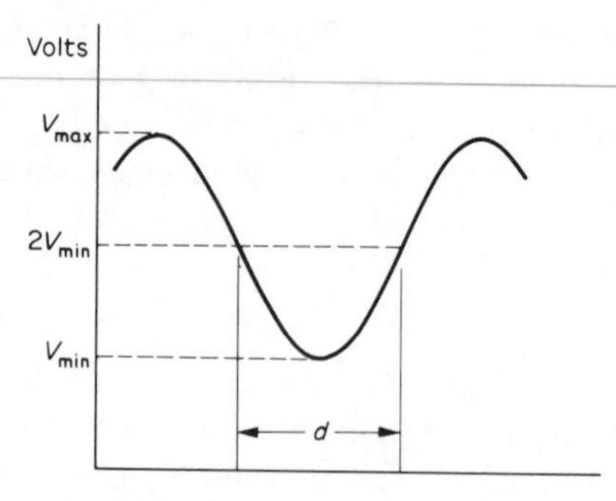

Fig. 51

VSWR is measured with a slotted section of waveguide and the output is fed into a high gain audio amplifier which is provided with a meter to read VSWR directly. By setting the probe at a maximum, the meter is adjusted to read unity and the probe is then moved to an adjacent minimum. The reading of the minimum can be directly calibrated in VSWR as it is the reciprocal of the actual meter reading.

Hence
$$\text{VSWR} = \frac{1}{V_{min}}$$

e.g. if $V_{min} = 0{\cdot}5\text{V}$, then $\text{VSWR} = \frac{1}{0{\cdot}5} = 2.0.$

High VSWR

For a VSWR > 10, the square law of the detector crystal in the probe is no longer true and the meter calibration would not be correct. In this case, the probe is set to a minimum on the scale and its position is noted. It is then moved to two other positions (one on either side of the minimum) where the meter reading is twice the minimum value. If d is the distance between these two positions as in Fig. 51, the VSWR is given by

$$\text{VSWR} = \frac{\lambda g}{\pi d}$$

Power measurement

The measurement of power in a microwave circuit is fundamental. Unlike voltage or current it is independent of the position in a waveguide and can be measured directly and conveniently by using thermally sensitive elements.

Power measurements are either at low, medium or high levels. Low levels (<10 mW) are measured by crystal rectifiers or bolometers, medium levels (<1 watt), by using directional couplers and attenuators with power meters.

The bolometer which is used as the sensing element is a device whose resistance varies in proportion to the r.f. energy incident on it, due to the conversion of r.f. energy into heat energy. Bolometers are either thin strands of platinum wire with a positive temperature coefficient and known as *barretters* or are small beads of semi-conducting material with a negative temperature coefficient and known as *thermistors*.

In the measurement of microwave power, a bolometer is mounted in a short section of a waveguide. The bolometer is connected as one arm of a power bridge which indicates power by measuring its resistance. The more accurate bridges usually maintain the bolometer resistance constant (about 100–200 ohm) by substituting a.c. (audio) power for the r.f. power.

Measurement of relative power levels can be accomplished by the detection of amplitude-modulated r.f. signals. The modulation is usually 1 kHz and the detected signal is amplified, rectified and drives a deflection meter. Since detection is square-law, the meter reading is proportional to the voltage squared or power. One such instrument which is calibrated directly using a square-law scale is the VSWR indicator.

For absolute power measurements, a bolometer must be properly mounted in a short section of waveguide so as to absorb all the incident power and efficiencies around 95% are possible. To eliminate errors, mismatches should be tuned out as an error of 0·1 dB is possible with a VSWR of 1·3.

EXAMPLE 14

Give the sequence of the component units of a waveguide test-bench suitable for an accurate measurement of the voltage standing-wave ratio in a section of waveguide operating at approximately 3 GHz. Describe the procedure and state the precautions necessary to avoid errors.

A standing-wave indicator is terminated in an unknown load and had a detector with a square-law response. If movement of the probe produces deflections which vary between 35 and 10% of full-scale, determine the reflection coefficient of the load.

(L.U.B.Sc(Eng.) Tels. Pt. 3, 1965)

Solution

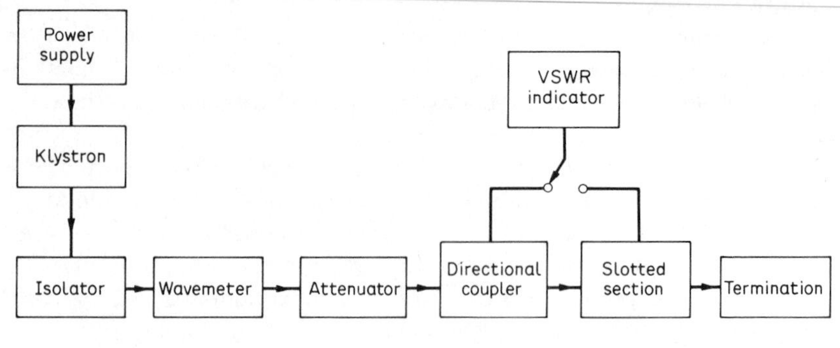

Fig. 52

Figure 52 shows a typical set-up for measuring VSWR. The reflex klystron is supplied with power from a stabilised power supply and is followed by an isolator which prevents any reflected power from reaching the klystron and thus ensures its stable operation. It is followed by a wavemeter for checking the frequency of the klystron and by an attenuator, which sets the level of power in the waveguide. The directional coupler then divides the power into two branches, one part being fed into a side-arm connected to a power meter, for monitoring the power output of the klystron, while the remainder passes down the slotted-section to the termination, which is either a short-circuit, open-circuit or a matched load. The S.W.R. indicator is connected to the output probe of the slotted section.

Procedure
The klystron supply is switched on and the repeller voltage is set to give a maximum output with 1 kHz square-wave modulation and the attenuator is adjusted to give a suitable deflection on the power meter. During the measurements, the power level on the power meter should be kept constant.

The wavemeter is now tuned to check the klystron at 3 GHz. At resonance, a drop in power level is indicated on the power meter and by tuning the klystron mechanically, the point of *minimum* dip will give the correct frequency required. The wavemeter is then completely detuned so as not to interact with the rest of the circuit.

The probe carriage over the slotted-section is now moved to and fro and the SWR indicator will show that maxima and minima exist over the slotted-section. To measure the VSWR, the method used depends on whether a low or high VSWR is present. Details were given in Section 6.3.

To avoid errors in the evaluation of the VSWR, the most sensitive scale should be used when measuring a low VSWR. When measuring a high VSWR, the position of a minimum is best determined by noting two positions on either side of it which give equal readings and the mean of these is the minimum reading. Furthermore, several readings should be taken of each measurement and the average obtained for the most accurate results.

PROBLEM

Since the crystal detector used in the probe is a square-law device, its output current is proportional to the *square* of the voltage in the waveguide and hence

$$\text{VSWR} = \frac{V_{\text{max}}}{V_{\text{min}}} = \sqrt{\frac{I_{\text{max}}}{I_{\text{min}}}} = \sqrt{\frac{100}{35}}$$

or VSWR $= 1{\cdot}69$

Problems

1 Explain the causes and effects of attenuation and phase distortion in audio frequency cables.

A cable is 20 miles long and has the following distributed constants per loop mile, $R = 40\,\Omega$, $L = 1$ mH, $C = 0{\cdot}06\,\mu$F. Shunt conductance can be neglected. Calculate the value of the characteristic impedance at 10 kHz. If the cable is terminated with this impedance, calculate for a 10 kHz signal (a) the wavelength and (b) the velocity of propagation. Calculate also the gain in dB of a repeater, inserted at the mid-point of the cable, to provide zero loss over the cable length.
(L.U.B.Sc(Eng) Tels. Pt. 3, 1963)

2 At 4 MHz, a coaxial cable has the following distributed constants per loop mile, $R = 110\,\Omega$, $L = 0{\cdot}455$ mH, $C = 0{\cdot}07\,\mu$F, and $G = 0{\cdot}003\,S$. Derive approximate expressions for the characteristic impedance and attenuation coefficient of the cable at this frequency.

The cable is used with intermediate repeaters in a multi-channel telephone system. If the attenuation at 4 MHz is not to exceed 50 dB between repeaters, calculate approximately the maximum repeater spacing allowable. (L.U.B.Sc(Eng) Tels. Pt. 3, 1964)

3 Explain what is meant by stub matching of high frequency transmission lines. What are the disadvantages of such a system?

A loss-free air-insulated transmission line, of characteristic impedance $50\,\Omega$, is connected to a resistive load of $120\,\Omega$. Derive expressions for (a) the position and (b) the length of a short-circuited stub that will terminate the main length of the line correctly. Hence determine the value of these quantities when the operating frequency is 85 MHz. The matching stub may be assumed to have a construction identical with that of the line, and the velocity can be taken as 3×10^8 m/s. (L.U.B.Sc(Eng) Tels. Pt. 3, 1968)

4 A section of waveguide terminated in an unknown impedance is connected to a waveguide test-bench. A VSWR of 4 is observed and minima are found when the detector probe is at 11·2 and 9·2 cm from the termination.

Mark, on the Smith Chart provided, points A and B corresponding to
(a) the impedance of the load.
(b) the admittance of the load.
Determine the shortest distance from the load at which insertion of a purely reactive shunt matching device will reduce the VSWR to unity.

State, with reasons, whether the matching device should be inductive or capacitive.

(C & G Adv. Mic. Principles, June 1969)

5 A coaxial line has a tubular inner conductor of radius 1 cm and an outer conductor of radius 5 cm, both conductors are of negligible thickness. The inner conductor is enclosed by a ferrite of constant thickness 1 cm and relative permeability 50. Calculate from first principles the inductance per metre. Determine also the stored energy per metre if a steady current of 1 A flows in one conductor and returns through the other; what fraction of this energy is stored in the ferrite? (L.U.B.Sc(Eng)El.Th. & Meas., Pt. 2, 1961)

6 State the conditions which must be satisfied by time-varying electromagnetic fields at the surface of a perfect conductor. Illustrate your answer by sketches of the electric and magnetic fields in the following special cases:
(a) a coaxial line,
(b) a pair of parallel lines remote from any other conductor,
(c) a waveguide.

(L.U.B.Sc(Eng) El. & Tels. Pt. 2, 1963)

7 A plane electromagnetic wave at a frequency of 2 GHz is travelling in air. The peak field intensity of the wave is 1 volt/m and it is incident normally on a very large sheet of copper. Calculate the penetration depth of the wave in the copper sheet and the power loss in the sheet per m^2 of surface area.

8 Discuss the propagation of an H_{10} wave in a rectangular waveguide. What determines the lowest frequency which can be transmitted along the guide?

The output of an oscillator is fed into a coaxial line and also into a waveguide. The coaxial line has air dielectric and losses are negligible. Probe measurements show that the distance between successive current nodes is 4 cm in the coaxial line and 5·4 cm in the waveguide.

Determine (a) the frequency of the oscillator (b) the phase velocity and (c) the group velocity, in the waveguide.

(L.U.B.Sc(Eng) Tels. Pt. 3, 1961)

9 Discuss the factors which affect the choice between coaxial cables and waveguides for UHF transmission and state the advantages and disadvantages of each.

What are the considerations which affect the choice of dimensions for a rectangular waveguide? Estimate suitable internal dimensions of a rectangular waveguide for the propagation of the fundamental H_{10} mode at a frequency of 10 GHz.

(L.U.B.Sc(Eng) Tels. Pt. 3, 1962)

10 Show how the concept of impedance can be generalised to apply to electromagnetic waves.

When $E_z = 0$, one of Maxwell's equations reduces to

$$-j\omega\mu H_y = \frac{\partial E_x}{\partial z}$$

Use this equation to show that the characteristic wave-impedance for an H-mode in a waveguide is given by $\omega\mu/\beta$, where β is the phase coefficient.

A closely fitting rectangular block of dielectric material of length l is inserted in a previously matched air-filled waveguide of rectangular cross-section, carrying an H-mode. Using the concept of wave-impedance, together with the transmission line formula

$$Z_s = Z_0\left[\frac{Z_R + jZ_0 \tan \beta l}{Z_0 + jZ_R \tan \beta l}\right]$$

show that the fraction $|\rho|^2$ of the incident power which is reflected by the dielectric block is given by

$$|\rho|^2 = \frac{(\beta^2 - \beta_0^2)^2 \sin^2 \beta l}{4\beta^2\beta_0^2 + (\beta^2 - \beta_0^2)^2 \sin^2 \beta l}$$

where β and β_0 are the phase coefficients of the dielectric-filled and air-filled sections of the waveguide respectively.

(L.U.B.Sc(Eng) Ad. El. Theory Pt. 3(U.C.) 1968)

11 A hollow metal tube is to be used as a microwave resonator. Develop an expression for the characteristic oscillation frequencies of the cavity. Define the Q factor and indicate how it may be calculated. What problems are associated with exciting the cavity?

(C.E.I. Pt. 2, Comm. Eng., October 1968)

12 Describe an experimental method for the accurate determination of *large* standing wave ratios which uses a moving-carriage standing-wave indicator with a square law detector. Measurements made in a slotted waveguide with such an instrument produced the following results:

θ	4	2	1	2	4
x	7·893	7·826	7·735	7·644	7·577

θ is the ratio of the observed output current of the square-law detector to that current as observed with the standing-wave indicator located at a position of a minimum deflection, and x is the reading of the vernier (in centimetres) which defines the position of the probe of the standing-wave indicator relative to a fixed reference plane.

Given that the guide wavelength was 4·00 cm, calculate the standing-wave ratio in the guide.

(L.U.B.Sc(Eng) Tels. Pt. 3, 1966)

13 (a) Describe the principle of a method of impedance measurement which is based on the use of the slotted-line, and discuss the design factors that determine the accuracy of measurement.

(b) A 75 Ω, solid-dielectric, slotted coaxial line impedance measuring equipment is connected to an unknown impedance. At a frequency of 200 MHz, a VSWR of 1·9 and a voltage antinode 32·2 cm from the terminals appear on the measuring line. Assuming the dielectric to be loss-free and to have a relative permittivity of 2·5, calculate the value of the unknown impedance.

(C & G Adv. Comm. Radio, June 1971)

14 A simple microwave test-bench consists of a reflex klystron, a 10 dB padding attenuator and a short-circuit load. Find the VSWR between the klystron and attenuator. Compare and contrast the performance with that of an isolator whose forward loss is 0·5 dB, backward rejection is 30 dB, put in place of the attenuator.

15 Sketch the lines of electric and magnetic fields for an E_{01} wave in a circular waveguide. The critical wavelength for such a wave is approximately 2·6r where r is the radius of the waveguide. Hence, or otherwise, develop the design of a piston attenuator for use at 2 GHz and comment on the range of attenuation that can usefully be achieved. How would the attenuator behave at higher and lower frequencies? (C.E.I. Pt. 2, Comm. Eng., 1970)

Answers

1 $141 \angle 16°$

$12{\cdot}4$ miles

$12{\cdot}4 \times 10^4$ miles/s

$25{\cdot}8$ dB

2 $Z_0 = L/C$

$\alpha = R/2Z_0 + GZ_0/2$

$7{\cdot}16$ miles

3 $0{\cdot}56$ m

$0{\cdot}48$ m

4 $0{\cdot}5$ cm

Capacitive

5 $7{\cdot}1\ \mu$H

$3{\cdot}55 \times 10^{-6}$ joules

$97{\cdot}6\%$

7 $1{\cdot}476 \times 10^{-6}$ m

$1{\cdot}65 \times 10^{-7}$ Watts/m^2

8 $3{\cdot}75$ GHz

$4{\cdot}1 \times 10^8$ m/s

$2{\cdot}2 \times 10^8$ m/s

9 $a = 4{\cdot}5 \times 10^{-2}$ cm

$b = 2{\cdot}25 \times 10^{-2}$ cm

$\lambda_c = 1{\cdot}5\lambda$ (assumed)

12 $7{\cdot}0$

13 $49{\cdot}5 + j\,31{\cdot}5\,\Omega$

14 *Attenuator:* Insertion loss = 10 dB, VSWR = $1{\cdot}22$

Isolator: Insertion loss = $0{\cdot}5$ dB, VSWR = $1{\cdot}07$

References

1 RAMO, S. and WHINNERY, J. R. *Fields and Waves in Modern Radio*. Wiley & Sons (1962). Chapter 6.

2 SHEPHERD, J., MORTON, A. H. and SPENCE, L. F. *Higher Electrical Engineering*. Pitman (1970).

3 STARR, A. T. *Telecommunications*. Pitman (1958). Chapter 5.

4 MORTON, A. H. *Advanced Electrical Engineering*. Pitman (1966).

5 GRIEG, D. D. and ENGELMANN, H. F. Microstrip—A new transmission technique for the kMHz range. *Proceedings Institute of Radio Engineers*, **40** (December 1952). Page 1644.

6 TRANSACTIONS INSTITUTE OF RADIO ENGINEERS. MTT-3, (March 1955). (Symposium.)

7 ASSADOURIAN F. and RIMAI, E. *Simplified Theory of Microstrip Transmission Systems. Proceedings Institute of Radio Engineers*, **40** (December 1952). Page 1651.

8 BARLOW, H. M. Guided electromagnetic waves. *The Radio and Electronic Engineer*, **41** (April 1971). Page 147.

9 INSTITUTE ELECTRICAL ENGINEERS. Conference Publication No. 71. *Trunk Telecommunications by Guided Waves*. London. (September 1970.)

10 PUPIN, M. I. *Wave transmission over non-uniform cables and long distance air-lines*, Trans. A.I.E.E. 17, p. 445 (1900).

11 CONNOR, F. R. *Signals*. Edward Arnold (1972).

12 SMITH, P. H. *Transmission Line Calculator*. *Electronics*. (January 1939), and *An Improved Transmission Line Calculator*, *Electronics*. (January 1944.)

13 RAMO, S. *et al*. *Fields and Waves in Communication Electronics*. Wiley & Sons (1965). Chapter 8.

14 STRATTON, J. A. *Electromagnetic Theory*. McGraw-Hill (1941). Chapter 1.

15 JORDAN, E. C. *Electromagnetic Waves and Radiating Systems*. Prentice-Hall Inc. (1968). Chapters 7 and 8.

16 RAMO, S. *et al*. *Fields and Waves in Communication Electronics*. Wiley & Sons (1965). Chapter 8.

17 SIMS, G. D. and STEPHENSON, I. M. *Microwave Tubes and Semiconductor Devices*. Blackie & Sons (1963).

18 TERMAN, F. E. *Electronic and Radio Engineering*. McGraw-Hill (1955).

19 LANCE, A. L. *Introduction to Microwave Theory and Measurements*. McGraw-Hill (1964).

20 REICH, H. J. *et al*. *Microwave Principles*. D. Van Nostrand Co. (1957). Chapter 10.

21 MONTGOMERY, C. G. *Techniques of Microwave Measurements*. Radiation Laboratory Series. McGraw-Hill (1947).

22 GINZTON, E. L. *Microwave Measurements*. McGraw-Hill (1957).

23 JORDAN, E. C. *Electromagnetic Waves and Radiating Systems*. Prentice-Hall Inc. (1968). Chapter 1.

24 RAMO, S. *et al*. *Fields and Waves in Communication Electronics*. Wiley & Sons (1965). Chapter 4.

Appendices

Appendix A: Vector analysis[23]

Scalar and vector quantities are used extensively in electromagnetic theory and a study of their properties is essential. A *scalar* is a quantity which has magnitude only while a *vector* is a quantity which has both magnitude and direction.

Unit vectors

To specify a vector, unit vectors are used. In Cartesian coordinates, these are **i**, **j**, **k** along the X, Y, Z axes respectively. Any general vector **V** with components x, y, z along the respective axes is given by $\mathbf{V} = \mathbf{i}x + \mathbf{j}y + \mathbf{k}z$.

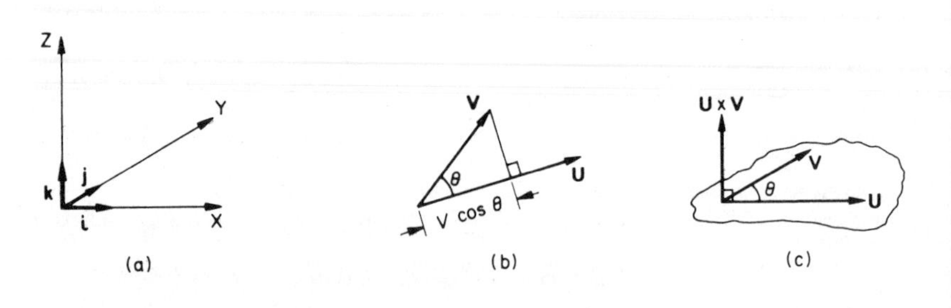

Fig. 53

Scalar product

The scalar or *Dot* product of **U** and **V** is written as **U . V** and is equal to the product of their magnitudes and the cosine of the angle between them. This is shown in Fig. 53.

$$\mathbf{U} \cdot \mathbf{V} = U_x V_x + U_y V_y + U_z V_z = UV \cos \theta$$

which is a scalar quantity.

Vector product

The vector or *Cross* product of **U** and **V** is written as $\mathbf{U} \times \mathbf{V}$. It is a vector whose magnitude is the product of the two vectors and the sine of the angle between them. The direction is normal to the plane containing **U** and **V** and is given by the right-hand screw rule. This is shown in Fig. 53.

$$\mathbf{U} \times \mathbf{V} = \mathbf{i}(U_y V_z - U_z V_y) + \mathbf{j}(U_z V_x - U_x V_z) + \mathbf{k}(U_x V_y - U_y V_x)$$

or

$$\mathbf{U} \times \mathbf{V} = \mathbf{n} UV \sin \theta$$

where **n** is the unit normal vector.

Gradient of a scalar

This is the maximum rate of change of a scalar function ϕ and conveys the idea of slope or *gradient*. It is a vector quantity and given by

$$\text{grad } \phi = \mathbf{i}\frac{\partial \phi}{\partial x} + \mathbf{j}\frac{\partial \phi}{\partial y} + \mathbf{k}\frac{\partial \phi}{\partial z}$$

in three dimensional space.

Divergence of a vector

The divergence of a vector **U** is a scalar quantity and it is a measure of the total outward flux of the vector, per *unit* volume.

$$\text{div } \mathbf{U} = \frac{\partial U_x}{\partial x} + \frac{\partial U_y}{\partial y} + \frac{\partial U_z}{\partial z} = \frac{1}{V}\int_S \mathbf{U} \cdot \mathbf{n} \, da$$

where V is the volume.

Curl of a vector

The curl of a vector is the line integral per unit area around an elementary area about a point, the area chosen is that which makes the line integral a maximum. It is also a vector and the positive direction is normal to the area and given by the right-hand screw rule. In determinant form it is

$$\text{curl } \mathbf{U} = \begin{vmatrix} \mathbf{i} & \mathbf{j} & \mathbf{k} \\ \dfrac{\partial}{\partial x} & \dfrac{\partial}{\partial y} & \dfrac{\partial}{\partial z} \\ U_x & U_y & U_z \end{vmatrix}$$

Operator ∇ (del)

This is a very useful operator and is defined by

$$\nabla \equiv \mathbf{i}\frac{\partial}{\partial x} + \mathbf{j}\frac{\partial}{\partial y} + \mathbf{k}\frac{\partial}{\partial z}$$

for Cartesian coordinates. Hence, we have

$$\text{grad } \phi = \nabla\phi$$
$$\text{div } \mathbf{U} = \nabla \cdot \mathbf{U}$$
$$\text{curl } \mathbf{U} = \nabla \times \mathbf{U}$$

Gauss' theorem

It relates a surface integral with a volume integral. This is shown in Fig. 54.

$$\int_S \mathbf{D} \cdot \mathbf{da} = \int_V \text{div } \mathbf{D} \, dv = \int_V \nabla \cdot \mathbf{D} \, dv$$

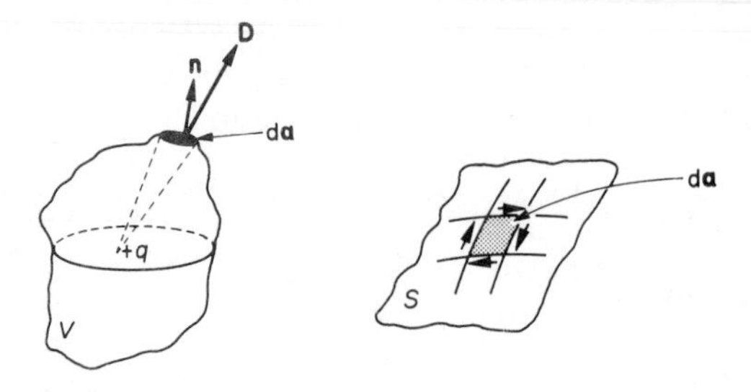

Fig. 54

Stokes theorem

It relates a line integral around a closed path with a surface integral. This is shown in Fig. 54.

$$\int_l \mathbf{B} \cdot \mathbf{dl} = \int_S \text{curl } \mathbf{B} \cdot \mathbf{da} = \int_S (\nabla \times \mathbf{B}) \cdot \mathbf{da}$$

Generalised coordinates

In some problems it is more convenient to use spherical or cylindrical coordinate systems are illustrated in Fig. 55.

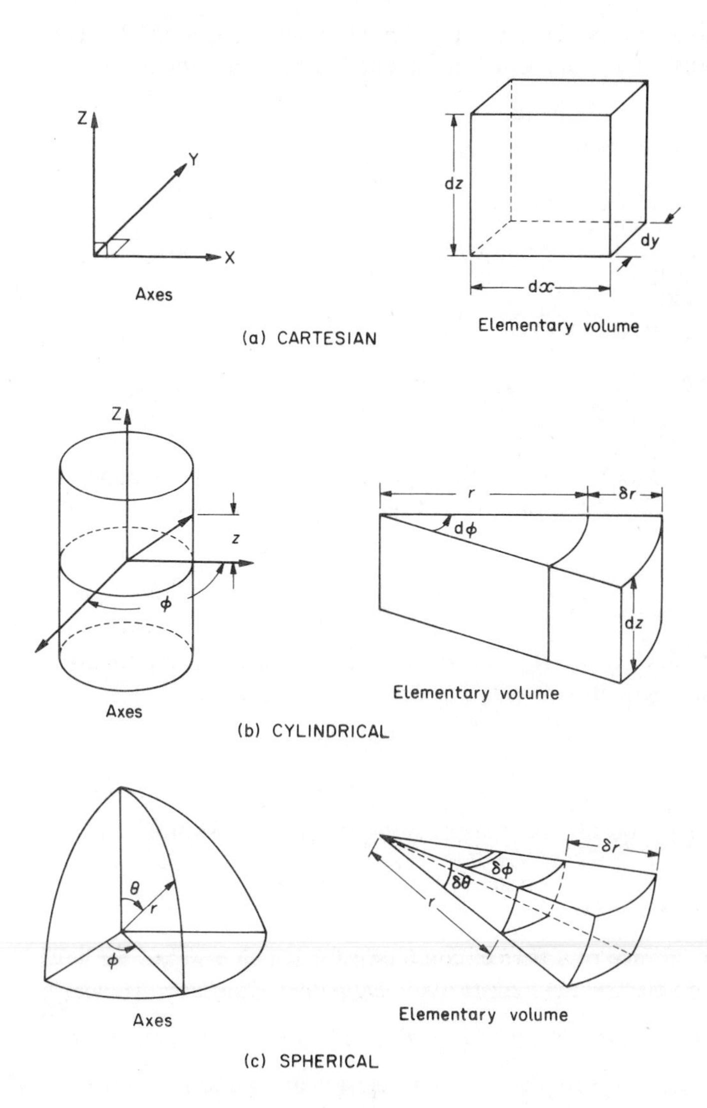

Fig. 55

Appendix B: Boundary conditions

(a) Boundary conditions[24]

For time-varying fields, certain boundary conditions exist at a conducting or dielectric surface which must satisfy Maxwell's equations.

Dielectric boundary

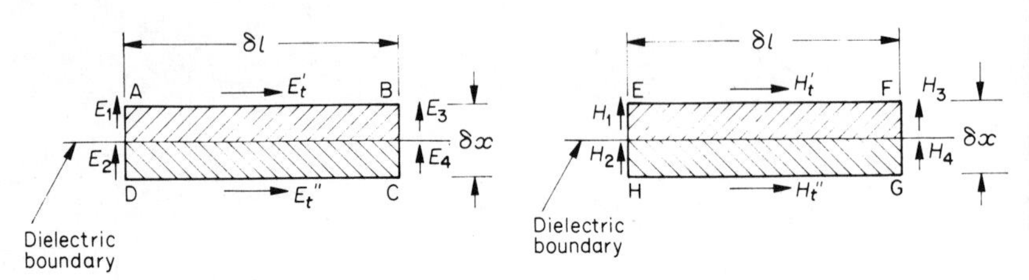

Fig. 56

If E'_t, E''_t and H'_t, H''_t are the tangential electric and magnetic components respectively, at the interface of the two media shown in Fig. 56, we must have

$$E'_t = E''_t$$

$$H'_t = H''_t$$

Similarly, if D'_n, D''_n, and B'_n, B''_n are the normal components of electric and magnetic flux density respectively, we must have

$$D'_n = D''_n$$

$$B'_n = B''_n$$

The proofs for these expressions are given in Ramo.[24]

Perfect conductor

At the surface of a perfect conductor for which $\sigma =$ o, we have $E''_t = 0$ since no electric field exists in an *equipotential* surface. Hence

$$E'_t = E''_t = 0$$

and so no tangential components exist near a perfect conductor. Normal components can exist as shown in Fig. 57.

Furthermore, as there are no magnetic poles at the surface, we have $B''_n = 0$.
Hence
$$B'_n = B''_n = 0$$
and so the magnetic lines form closed loops as shown in Fig. 57.

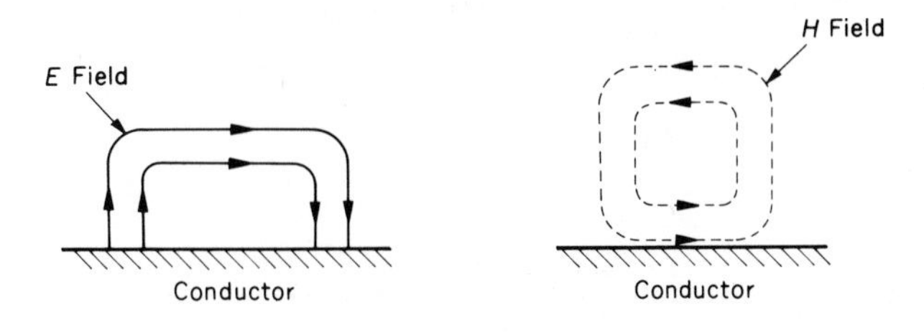

Fig. 57

The boundary conditions are useful in determining the field patterns inside bounded regions such as waveguides.

(b) Surface reflections

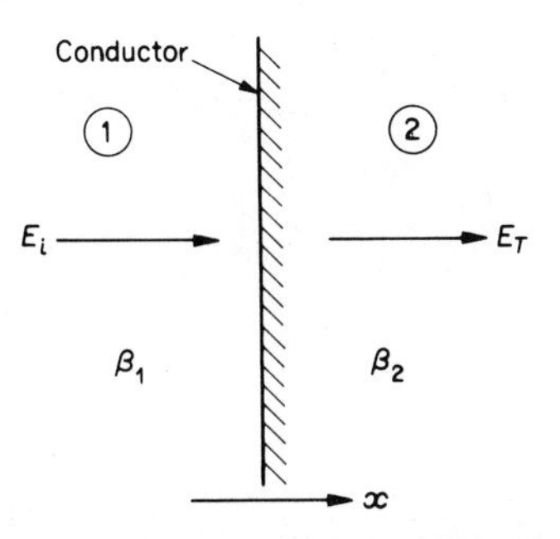

Fig. 58

Consider a wave incident normally at the conductor boundary in Fig. 58. Let the incident wave be denoted by
$$E_i = E_1 e^{j(\omega t - \beta_1 x)}$$
and the transmitted wave by
$$E_T = E_2 e^{j(\omega t - \beta_2 x)}$$
where β_1 and β_2 are the phase-change coefficients outside and inside the conductor respectively.

In general the phase change coefficient β for a conducting medium is given by*

$$\beta = \sqrt{\omega^2 \mu\varepsilon - j\sigma\omega\mu}$$

where σ is the conductivity.

For metals $$j\sigma\omega\mu \gg \omega^2 \mu\varepsilon$$

Hence $$\beta_2 \simeq \sqrt{-j\sigma\omega\mu}$$

$$\sqrt{\frac{\sigma\omega\mu}{2}} \cdot \sqrt{-2j}$$

$$\simeq \sqrt{\frac{\sigma\omega\mu}{2}}(1 - j)$$

or $$\beta_2 = s(1 - j)$$

where $$s = \sqrt{\frac{\sigma\omega\mu}{2}}$$

Hence $$E_T = E_2 e^{(j\omega t - jsx - sx)} = E_2 e^{-sx} \,.\, e^{j(\omega t - sx)}$$

The penetration or skin depth δ is defined such that the amplitude E_T in the conductor is reduced to $1/e$ of its initial value E_2. This happens when $sx = 1$ with $x = \delta$.

Hence $$s\delta = 1$$

or $$\delta = \sqrt{\frac{2}{\sigma\omega\mu}}$$

Appendix C: Cavity resonators

Fields can exist in regions entirely bounded by conducting walls called cavities. The cavities may be rectangular or circular in shape and behave very much like resonant circuits. If a small loop is coupled into the cavity and excited by an oscillator, very large fields can exist in the cavity at certain frequencies only. The general principle is that standing waves are set up in the cavity which satisfy the boundary conditions at the specific frequencies, if the cavity dimensions are correct.

* See Section 4.2(b).

Q of a cavity

The Q of an inductance or capacitance is a measure of its ability to act as a pure reactance. There will be dielectric or ohmic losses that will cause a small energy loss every time energy is stored in the reactance. For a coil of inductance L in series with a resistance R_s, the Q is given

$$Q = \omega_0 L/R_s$$

where ω_0 is the resonant angular frequency

or $$Q = \omega_0 \cdot \frac{\frac{1}{2}LI_m^2}{\frac{1}{2}I_m^2 R_s} = \omega_0 \cdot \frac{\text{(peak energy stored)}}{\text{(average energy loss)}}$$

This latter definition is more appropriate for cavities where $Q = Q_U$, the *unloaded* Q of the cavity if the average energy loss occurs in the cavity only. Denoting peak stored energy as U and average loss as W, we have

$$U = \int_V \frac{\varepsilon E_m^2}{2}\,dv = \int_V \frac{\mu H_m^2}{2}\,dv$$

If there are no dielectric losses but only ohmic losses in the cavity walls, we have

$$W = \int_S \frac{|H_t|^2 R_s}{2}\,da$$

where $|H_t|$ is the tangential magnetic field component which numerically equals the surface current and R_s is the surface resistivity.

Now $$R_s = 1/\sigma\delta$$

where σ is the conductivity and δ is the penetration depth given by

$$\delta = \sqrt{\frac{2}{\omega_0 \mu\sigma}}$$

Hence $$Q_U = \frac{\omega_0\mu\sigma\delta \int_V H^2\,dv}{\int_S H^2\,da} = 2/\delta \frac{\int_V H^2\,dv}{\int_S H^2\,da}$$

$$Q_U = 2/\delta \frac{\text{(mean of } H^2 \text{ in the volume} \times \text{volume)}}{\text{(mean of } H^2 \text{ over the surface} \times \text{surface)}}$$

with $$Q_U \simeq 2/\delta \,.\, \frac{\text{Volume}}{\text{Area}}$$

if the mean of H^2 in the volume and over the surface are approximately equal for the cavity.

If the cavity is coupled to a load as in Fig. 59, the loaded Q is given by Q_L where

$$Q_L = \omega_0 \cdot \frac{\text{(stored energy)}}{\text{energy loss in cavity \& load}}$$

Hence

$$\frac{1}{Q_L} = \frac{\text{energy loss in cavity}}{\omega_0 \times \text{stored energy}} + \frac{\text{energy loss in load}}{\omega_0 \times \text{stored energy}}$$

or

$$\frac{1}{Q_L} = \frac{1}{Q_U} + \frac{1}{Q_{\text{Load}}}$$

which means that the loaded Q of the cavity is less than its unloaded value. Energy may be coupled into and out of the cavity by means of loops, as in Fig. 59.

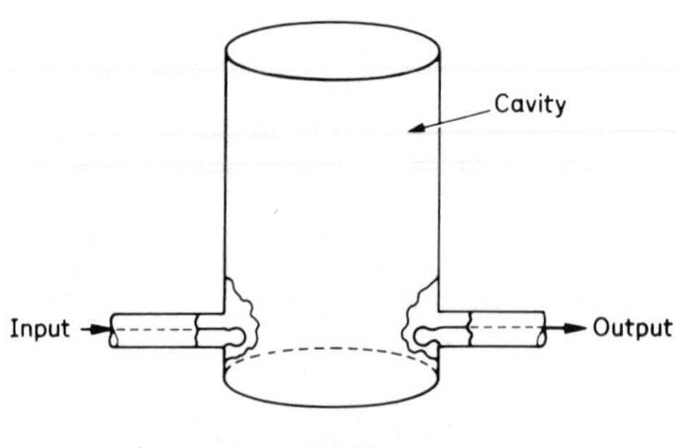

Fig. 59

Cylindrical cavities are commonly used in practice as they have a large volume to surface area which gives a high value of Q_U.

Cavity modes

These are basically waveguide modes since a resonant cavity can be constructed by shorting a length of waveguide (rectangular or circular) with metal end plates. A standing wave pattern will exist in the cavity provided the boundary conditions at the metal walls are satisfied. This is achieved if the cavity length l is an integral number of $\lambda g/2$.

Cavity modes are specified by the two basic subscripts m, n used for rectangular waveguides. In addition, a third subscript p or q designates the number of half wavelengths along the cavity length l. Hence, we have $TE_{mnp}(H_{mnp})$ modes or $TM_{mnp}(E_{mnp})$ modes for rectangular cavities and $TE_{mnq}(H_{mnq})$ modes or $TM_{mnq}(E_{mnq})$ modes for circular cavities.

Rectangular cavities

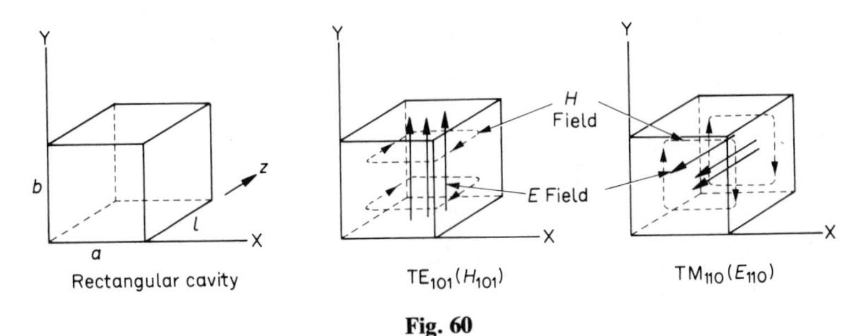

Fig. 60

Let the cavity dimensions be a, b, l as shown in Fig. 60. The additional boundary condition to be satisfied is given by

$$l = p\lambda_g/2$$

where $p = 0, 1, 2$ etc. and the condition $p = 0$ implies a mode in which the electric field lines are normal to the end walls, in the z-direction.

Since the cavity mode is basically a waveguide mode, we have

$$1/\lambda^2 = 1/\lambda_c^2 + 1/\lambda_g^2$$

with

$$\lambda_g = 2l/p$$

Since

$$\lambda_c = \frac{2}{\sqrt{(m/a)^2 + (n/b)^2}}$$

the resonant wavelength λ of the cavity is given by

$$1/\lambda^2 = \frac{(m/a)^2 + (n/b)^2}{4} + p^2/4l^2$$

or

$$\lambda = \frac{2}{\sqrt{(m/a)^2 + (n/b)^2 + (p/l)^2}}$$

for TE_{mnp} or TM_{mnp} modes.

Cylindrical cavities

These cavities are easier to construct mechanically with a high Q and are greatly used in practice. The cavity modes are designated TE_{mnq} or TM_{mnq} where q refers to the number of $\lambda_g/2$ along the length l indicated in Fig. 61.

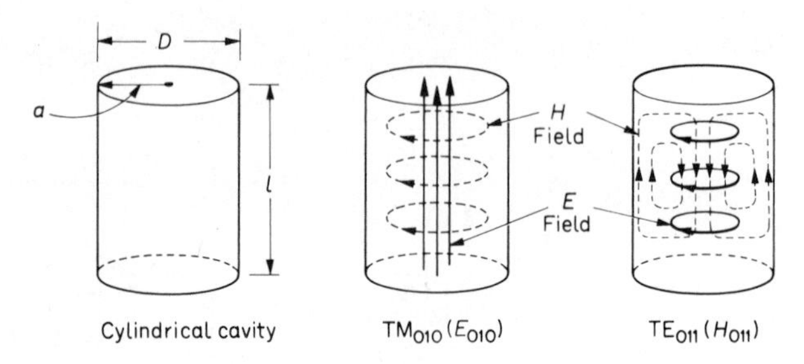

Fig. 61

If λ is the resonant wavelength for a cavity of length l and radius a, we have for TE modes

$$1/\lambda^2 = 1/\lambda_c^2 + 1/\lambda_g^2$$

with

$$l = q\lambda_g/2$$

where $q = 0, 1, 2$ etc. and $\lambda_c = 2\pi/k_c$.

Since $k_c = p'_{mn}/a$ where p'_{mn} is the nth root of the equation $J'_m(k_c a) = 0$, we obtain

$$1/\lambda^2 = (k_c/2\pi)^2 + (q/2l)^2$$

or

$$1/\lambda^2 = \left(\frac{p'_{mn}}{2\pi a}\right)^2 + (q/2l)^2$$

If $D = 2a$ is the cavity diameter, then

$$1/\lambda^2 = (p'_{mn}/\pi D)^2 + (q/2l)^2$$

or

$$\lambda = \frac{1}{\sqrt{(q/2l)^2 + (p'_{mn}/\pi D)^2}} \qquad \text{for TE modes}$$

and by similar reasoning p'_{mn} is replaced by p_{mn} for TM modes and we obtain

$$\lambda = \frac{1}{\sqrt{(q/2l)^2 + (p_{mn}/\pi D)^2}} \qquad \text{for TM modes}$$

A very useful mode employed for wavemeters is the $TE_{011}(H_{011})$ shown in Fig. 61.

The electric field lines are loops in the centre while the magnetic lines travel through them and down by the side walls. Hence, no currents flow between the end plate and walls. This is useful in a tunable cavity since movement of an end plate as a piston, does not disturb the field pattern. As there is a one half-wavelength variation of E along l, $q = 1$ with $p'_{mn} = p'_{01} = 3{\cdot}83$ and we obtain

$$\lambda = \frac{1}{\sqrt{(1/2l)^2 + \left(\dfrac{1}{0{\cdot}82D}\right)^2}}$$

as the resonant wavelength of the cavity.

Index

9.12.85